Green Energy and Technology

For further volumes:
http://www.springer.com/series/8059

Takeshi Yao
Editor

Zero-Carbon Energy Kyoto 2012

Special Edition of the Joint Symposium "Energy Science in the Age of Global Warming" of the Kyoto University Global COE Program and the JGSEE/CEE-KMUTT

Springer

Editor
Takeshi Yao
Professor of the Graduate School of Energy Science
Kyoto University
Steering Committee of GCOE Unit for Energy Science Education
Yoshida-honmachi, Sakyo-ku
Kyoto 606-8501, Japan
gcoe-office@energy.kyoto-u.ac.jp

ISSN 1865-3529 ISSN 1865-3537 (electronic)
ISBN 978-4-431-54263-6 ISBN 978-4-431-54264-3 (eBook)
DOI 10.1007/978-4-431-54264-3
Springer Tokyo Heidelberg New York Dordrecht London

Library of Congress Control Number: 2013932569

Printed on acid-free paper

Springer is part of Springer Science+Business Media (www.springer.com)

Preface

The Fourth International Symposium of the Global COE (Center of Excellence), titled "Zero-Carbon Energy, Kyoto 2012," was held jointly with the Joint Graduate School of Energy and Environment/Center of Excellence on Energy Technology and Environment (JGSEE/CEE) at King Mongkut's University of Technology, Thonburi (KMUTT) in Bangkok, Thailand, on May 22–23, 2012, succeeding the International Symposium "Zero-Carbon Energy, Kyoto 2009" at Kyoto University Clock Tower, "Zero-Carbon Energy, Kyoto 2010" at Kyoto University Obaku Plaza, and "Zero-Carbon Energy, Kyoto 2011" at Suwon, Korea. This 2012 symposium provided an opportunity for researchers to present their scenarios and their advanced research works. Many important lectures and discussions by invited speakers and members of the Global COE, as well as interesting presentations by students of the GCOE Unit for Energy Science Education, were given. This book is a compilation of the lectures and presentations.

This is the final year of the 5-year Global COE Program "Energy Science in the Age of Global Warming—Toward a CO2 Zero-Emission Energy System" of the Ministry of Education, Culture, Sports, Science and Technology of Japan. The program has aimed to establish an international education and research platform to foster educators, researchers, and policy makers who can develop technologies and propose policies for establishing a scenario for a CO2 zero-emission society no longer dependent on fossil fuels by the year 2100. Since 2008, four departments of Kyoto University—the Graduate School of Energy Science, the Institute of Advanced Energy, the Department of Nuclear Engineering, and the Research Reactor Institute—have joined together and have been engaged in the management of the Global COE Program.

The Global COE has established the GCOE Unit for Energy Science Education at its center and has placed the Scenario Planning Group, the Advanced Research Cluster, and the Evaluation around the Unit forming mutual associations. The Scenario Planning Group has set out CO2 zero-emission technology roadmaps and established CO2 zero-emission scenarios, with analyses from the standpoints of social values and human behavior. The group has set up the Committee of

Energy Scenario and Strategy Study to cooperate with the government and industries on energy and environmental issues. The Advanced Research Cluster, as an education platform based on research, has promoted a socio-economic study of energy, a study of new technologies for solar energy and biomass energy, and research for advanced nuclear energy following the road map established by the Scenario Planning Group. The Cluster has set up the Joint Committee of Scenario Planning and Advanced Research in order to enhance cooperation between the Scenario Planning Group and the Advanced Research Cluster.

At the GCOE Unit for Energy Science Education, the students have planned and conducted interdisciplinary group research containing both the social and human sciences as well as the natural sciences toward the goal of CO2 zero emission at the initiative of the students themselves. The students have acquired the ability to survey the whole energy system through participation in scenario planning and interaction with researchers from other fields and have applied it to their own research. This approach is the major feature of the cultivation of human resources. The Global COE has been striving to foster young researchers who will be able to employ their skills and knowledge with a broad international perspective and expertise in their fields of study in order to respond to the needs of society in terms of diverse energy and environmental problems.

In order to transmit the achievements of this platform to the public, the Global COE has posted information on a web site and has published annual reports, quarterly newsletters, books, and self-inspection and evaluation reports. It has also hosted domestic and international symposiums and the GCOE industry-government-academia collaboration symposium and citizen lectures, and has co-hosted related meetings both domestically and internationally.

Securing energy and conservation of the environment are the most important issues for the sustainable development of human beings. The energy problem cannot be simply labeled a technological one, as it is also deeply involved with social and economic elements. It is necessary to establish low carbon-energy science as an interdisciplinary field, bringing together the social and human sciences and the natural sciences.

Takeshi Yao
Program Leader
Global COE "Energy Science in the Age of Global Warming
—Toward a CO_2 Zero-emission Energy System"

Contents

Part I
Scenario Planning and Socio-economic Energy Research

(i)
Invited Paper

Challenges of Nuclear Safety to Sustainable Development of Chinese Nuclear Energy in Post-Fukushima Era

Zhiwei Zhou

Abstract Sustainable development of Chinese economy in twenty-first century will mainly rely on self-supply of clean energy with indigenous natural resources. The burden of current coal-dominant energy mix and the stress on reduction of CO_2 emission due to energy consumptions has led nuclear power to be an indispensable choice for further growth of Chinese electricity generation capacity. This paper describes the challenges nuclear safety issues in post-Fukushima era related to the development of nuclear industry of China. Three major challenges that China's nuclear energy market is facing are addressed, which include: safety standards, speed of NPP construction, and advanced technologies. This study emphasizes that China should implement sustainable energy development policy and pay great attention to advanced nuclear technologies with passive and inherent safety in future.

Keywords Advanced nuclear technology • Chinese nuclear policy • Fukushima accident • Nuclear safety • Sustainable development

1 Introduction

The time span of next 15–20 years is a crucial period for China to develop prosperous economy and to establish harmonic society in all aspects benefiting its well over 1 billion people. Energy is one of the most important material conditions. The bottle-neck burdens from energy resources and environmental protection will be the major limits for fulfilling the criteria of a modern harmonic and prosperous society.

Z. Zhou (✉)
Institute of Nuclear and New Energy Technology of Tsinghua University, Beijing, China
e-mail: zhouzhw@mail.tsinghua.edu.cn

T. Yao (ed.), *Zero-Carbon Energy Kyoto 2012*, Green Energy and Technology,
DOI 10.1007/978-4-431-54264-3_1,

To realize harmonic between man and nature is the up-most ambition of human-being. The rich culture embedded in China's long lasting civilization of more than 3,000 years has always been reflecting the thoughts of "harmonic combination between man and nature" from numerous great ideologists, politicians and scientists. The modern industrial technology based on western scientific civilization also pursues the philosophy of sustainable development. Nowadays, more than 400 nuclear power plants are producing safely and reliably about 14% of total electricity used by human being without emission of greenhouse gas CO_2 [1]. This is a great achievement made by nuclear energy technology which originated from the middle of the twentieth century.

Although many challenges have resulted in a valley in projecting nuclear power plants under construction worldwide in the previous two decades, nuclear energy so far is still a competitive alternative option to substitute fossil energy in massive scale for generating electricity in predictable near future. The major trend in last a few years in the world for new-built nuclear power plants (NPP) projects before Fukushima accident showed a strong indication of the renaissance of nuclear energy. According to the forecast from the prestigious research institutions [2], the potential market for nuclear power plants in China is in the scale of 240–400 GWe. However, current total generation capacity of nuclear power plants (NPPs) of China is only about 11 GWe. Now, China has played the leading role to accelerate the worldwide nuclear renaissance and 28 units of NPP are under construction. Unfortunately, the Fukushima accident induced by the devastating tsunami after a beyond design basis earthquake may have some setback impacts to the development of global nuclear energy industry. The global statistic poll has shown that the public support rate for continuing nuclear energy still remains high, although it drops in comparison to the situation before Fukushima accident. So far, China's policy to effectively develop nuclear energy on the basis of safety first principle has not changed. It will be foreseeable that the potential Chinese market for new units of NPPs will still remain very attractive to both international and domestic vendors and equipment suppliers.

2 Challenges in China's Energy Security and Nuclear Safety

2.1 China's Energy Consumption and Supply

The primary energy consumption of China since 2006 has been listed in Table 1. The data shown in table one indicates that China has become the largest energy consuming country in the world since 2009, although the level of the primary energy consumption per capita in China (~2.5 tce) is still very low compared to the world rich countries, such as USA (~8 tce), Japan (~4 tce). However, the total energy production of China has almost matched up the energy consumption in

Table 1 Total energy consumption of China since 2006

Year	Total consumption (billion tce)
2006	2.46
2007	2.65
2008	2.8
2009	3.06
2010	3.25
2011	3.48
2020 (forecast)	~4.0 (coal share >60%)

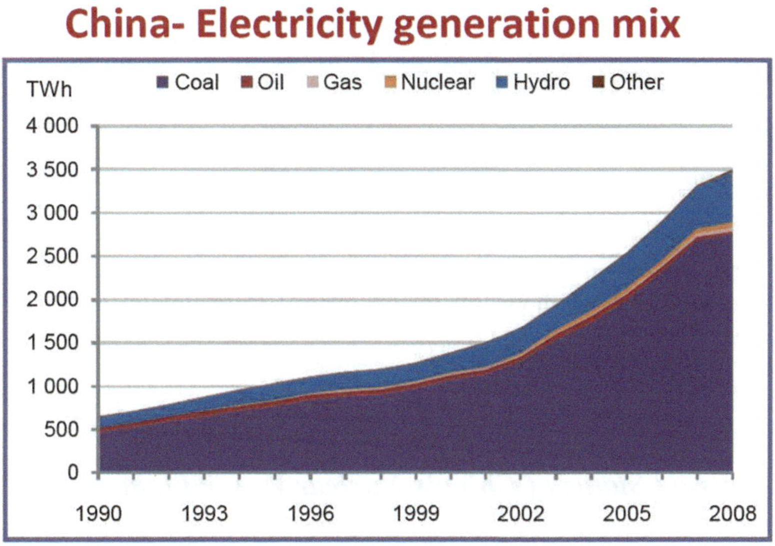

Fig. 1 China's electricity generation mix

China and therefore China is also the largest energy producer. The main contributors of energy sources so far in China are the coal (~70%) and hydropower (~16%). Thus, the room for developing new energy, such as wind, solar, biomass and other types of renewable energy, and as well nuclear energy, remains very large (see Figs. 1 and 2).

The general principle of energy policy of China is to insist on self-reliance to ensure energy security. The current status is that self-supply occupies about 94% share of Chinese total energy market and the net imported energy shares only 6% of total demand of Chinese energy market. The major part of the imported energy has been crude oil and natural gas, and the uranium import may also increase in near future. It is projected that the above proportional ratio of Chinese energy consumption and supply will not significantly change in the decade to come.

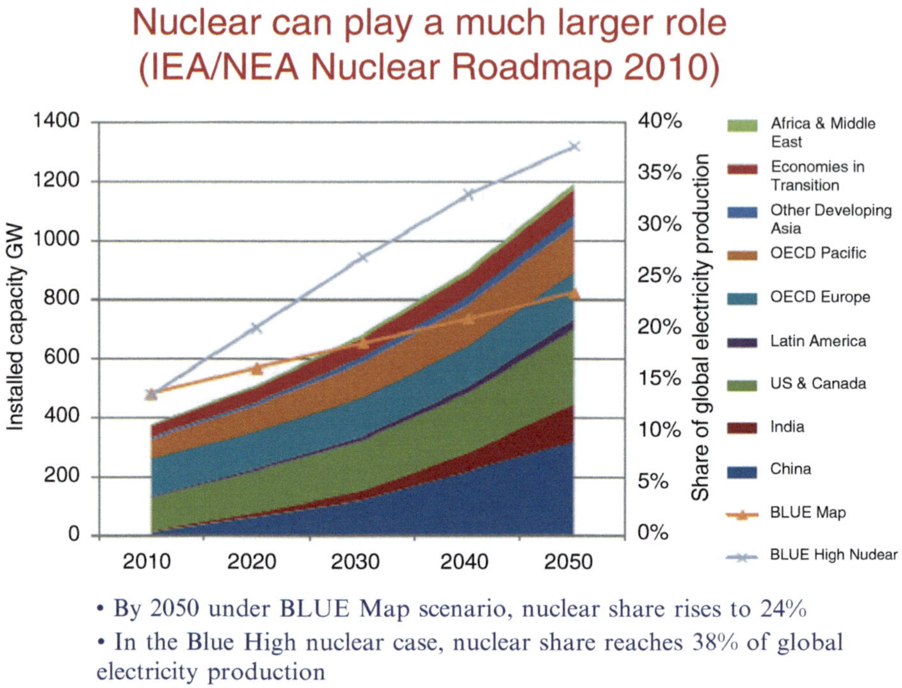

Fig. 2 Potential role of nuclear generation capacity

2.2 Current Status of China's Nuclear Power Plants

Nuclear power plants in commercial operation in China and their operational performances are listed in Table 2, indicating all NPPs of China in commercial operation are very reliable.

Nuclear power plants under construction in China and the type of their reactors are listed in Table 3, in which total of 28 units include various types of PWRs and modular high temperature reactors and cover GEN-II+, GEN-III and GEN-III+ reactors. Table 3 has also shown that Chinese nuclear fleet is experiencing transition from GEN-II+ to more advanced GEN-III or plus technologies. Safety has always been emphasized as the priority of the new-built NPPs.

Figure 3 indicates the distribution of existing and planned NPPs sites in China. The future trend is to expand NPP constructions in a number of inland sites. From the global experience, the inland NPP sites are significantly more than the coastal sites because it is much easier to find the low population density sites which are satisfy the criteria of building NPPs in the vast inland area.

China is a continental country, the remote northern and north-east areas along the border to Mongolia and Russia may be suitable for building new inland NPPs, by taking into account the easy distance for transmitting electricity to the industrial

Table 2 Operational performance of Chinese nuclear power plants

NPP name	Type of reactor	Capacity factor	Load factor
Qinshan	PWR 300 MWe, GEN-II	96.39%	95.58%
Daya Bay	U-1 PWR 900 MWe, GEN-II+	99.78%	99.60%
	U-2 PWR 900 MWe, GEN-II+	86.20%	86.39%
Qinshan-II	U-1 PWR 600 MWe, GEN-II+	85.35%	87.41%
	U-2 PWR 600 MWe, GEN-II+	85.21%	87.00%
Lingao	U-1 PWR 1 GWe, GEN-II+	92.11%	90.72%
	U-2 PWR 1 GWe, GEN-II+	85.24%	84.57%
Qinshan-III	U-1 CAN 650 MWe, GEN-II	91.21%	93.52%
	U-2 CAN 650 MWe, GEN-II	87.32%	89.25%
Tianwan	U-1 VVER 1 GWe, GEN-II+	70.97%	74.76%
	U-2 VVER 1 GWe, GEN-II+	81.20%	85.47%
Lingao-II[a]	U-1PWR 1 GWe, GEN-II+	n/a	n/a
	U-2 PWR 1 GWe, GEN-II+	n/a	n/a
Qinshan-II[a]	U-3,4 PWR 600 MWe, GEN-II+	n/a	n/a
	Average factor	87.36%	88.57%

[a]NPPs start to operate in 2010, the operation data are not available for a year based average

Table 3 NPPs under construction in China

Contractor	Site location	Type of reactor	Units and power
SNPTC	Sanmen, Zejiang	AP1000, GEN-III	2 × 1000 MWe
	Haiyang, Shandong	AP1000, GEN-III	2 × 1000 MWe
CGNPC	Taisan, Guangdong	EPR, GEN-II+	2 × 1700 MWe
	Hongyanhe, Liaoning	CPR1000, GEN-II+	4 × 1000 MWe
	Ningde, Fujian	CPR1000, GEN-II+	4 × 1000 MWe
	Yangjiang, Guangdong	CPR1000, GEN-II+	3 × 1000 MWe
	Fangchenggang, Guangdong	CPR1000, GEN-II+	2 × 1000 MWe
CNNC	Fuqing, Fujian	M310+, GEN-II+	3 × 1000 MWe
	Fangjiashan, Zejiang	M310+, GEN-II+	2 × 1000 MWe
	Changjiang, Hainan	CNP600, GEN-II+	2 × 600 MWe
Huaneng	Shidaowan, Shandong	HTR-PM, GEN-III+	2 × 100 MWe

and commercial prosperous flat and coastal zones in north-east provinces and municipals. Other inland areas in central and south-west provinces are also possible to find suitable inland sites for NPP if the domestic economic boost reaches these areas.

2.3 China's Nuclear Safety Regulatory Body

China has established a legal system to regulate and oversight nuclear safety. The safety standards developed by International Atomic Energy Agency (IAEA) have been adopted as the backbone in Chinese nuclear safety regulation system.

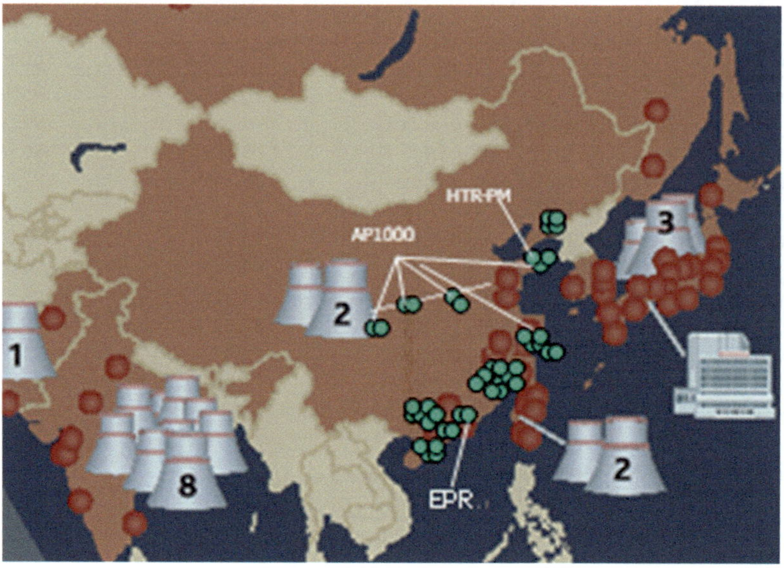

Fig. 3 Existing and planned sites of China's NPP

Meanwhile, a professional team has also been established for performing safety assessment and inspection to nuclear power plants and the other civil nuclear facilities. The Ministry of Environmental Protection (MEP) and the National Nuclear Safety Administration (NNSA) under MEP have strengthened regional offices which are located at Beijing, Shanghai, Shenzhen, Chengdu, Dalian and Lanzhou. As the technical support centers, Nuclear and Radiation Safety Center, Radiation Monitoring Center, Mechanical Equipment Reliability Center, Suzhou Safety Center and Beijing Nuclear Safety Review Center have played vital roles in nuclear safety assessment. Among them, Nuclear and Radiation Safety Center are of full responsibility to carry out safety assessment for NPP license. Figure 4 displays the institutional structure of the NNSA.

The main measures for safety regulatory body of China to ensure NPP safety include: adopting the "safety first" policy firmly; enhancing the legislation and regulation on nuclear safety; ensuring nuclear safety to the new projects; strengthening nuclear supervision and surveillance; establishing high standards for civil nuclear-grade equipments; increasing government funding for supervision staff; enhancing nuclear emergency readiness and training; implementing periodic safety review and experience feedback; applying probability safety assessment (PSA) and strengthening scientific research on severe accident management; enhancing international cooperation in the field of nuclear safety; and advocating the information publicizing and encouraging public participation positively.

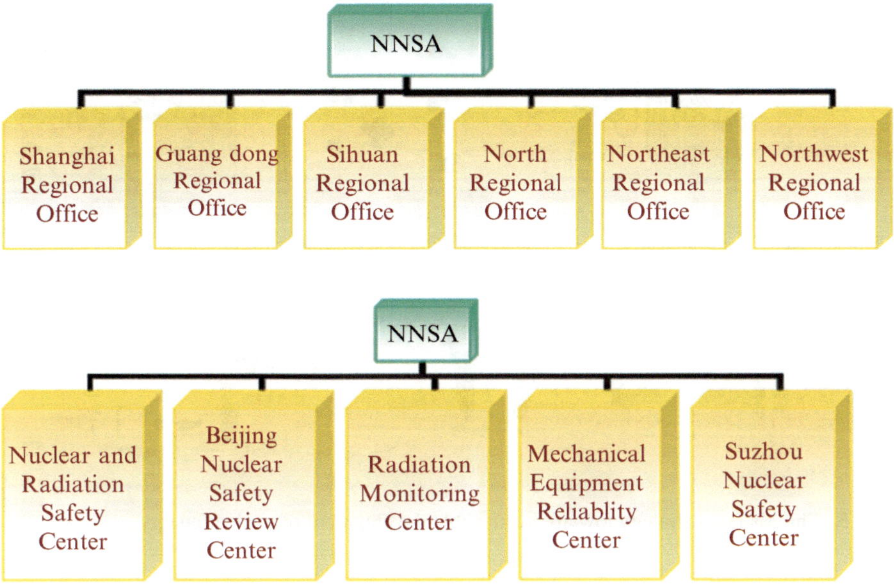

Fig. 4 Institutional structure of NNSA

2.4 *Post-Fukushima Actions*

Since March 11 when the Fukushima-I NPP was struck by the earthquake and the following 14 m high tsunami [3, 4], Chinese government has paid great attention to the evolving of the Fukushima accident. When the accident became a severe accident with hydrogen explosions, Chinese government immediately took a firm decision to assess the safety status of all NPPs both in operation and in construction, and to suspend all applications for constructing new nuclear power plants temporarily. The following major actions have been implemented: (1) an expert group has been assigned to complete specified nuclear safety inspection to all NPPs in operation to evaluate the safety status. If any one could not meet the safety requirements, it should be shutdown for further evaluation; (2) all NPPs under constructions has been inspected by the expert task force to ensure new NPP construction to fulfill safety requirements and high quality standards; (3) all NPPs planned to construct must adopt advanced safety standards; (4) all new applications for new NPP projects have been temporarily suspended until the new nuclear safety plan (up to 2020) be approved.

It has been clear that "Safety first, quality assurance, advanced technology, sustainable development" become the main policy to future NPP projects. However, the policy to develop nuclear energy efficiently based on "safety first" principle still remain unchanged.

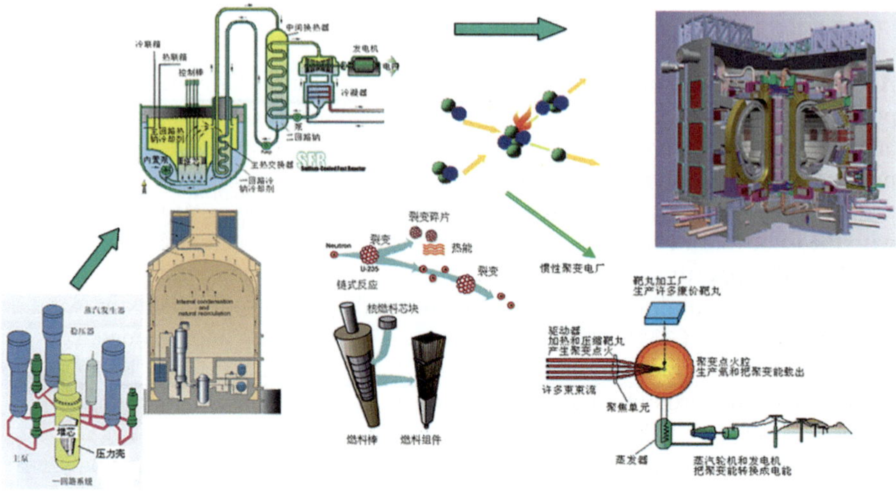

Fig. 5 China's main strategy towards sustainable development of nuclear energy

2.5 *Future Development Strategy of China's Nuclear Energy*

The sustainable development of Chinese economy will still rely on cheap, clean, low carbon and reliable energy source [5]. No matter we like it or not, nuclear energy is still the only alternative to substitute the fossil energy in massive scale. However, with the implementing of advanced safety standards, more advanced NPP technologies, may become the backbone of China's future NPP industry. The consensus on the main strategy to develop China's sustainable nuclear energy has been achieved in China's nuclear community, and the roadmap can be briefed as "PWR-FBR-Fusion (Pressurized Water Reactor – Fast Breeder Reactor – Fusion reactor)" as shown in Fig. 5.

It is clear that in the near future before the economical fast breed reactors and fusion reactors are attainable, the GEN-III PWRs will play the major role. Two types of GEN-III PWRs currently in construction in China, namely the EPR with active safety characteristics and the AP1000 with passive safety features, are illustrated in Figs. 6 and 7. The generation III PWR has at least one order higher safety standard evaluated by core damage frequency (CDF $< 10^{-5}$/a) and large radioactive release frequency (LRF $< 10^{-6}$/a) than generation II PWR. Especially, the passively safe PWR-AP1000 has achieved very low CDF ($<10^{-6}$/a) and LRF ($<10^{-7}$/a).

3 Security of NPPs and Other Nuclear Facilities in China

Security of NPP in China has been addressed in order to prevent any NPP from sabotage. In the aspect of the physical protection to NPPs, the emphasis has focused on entrance control, protection with armed police; regarding the issue of

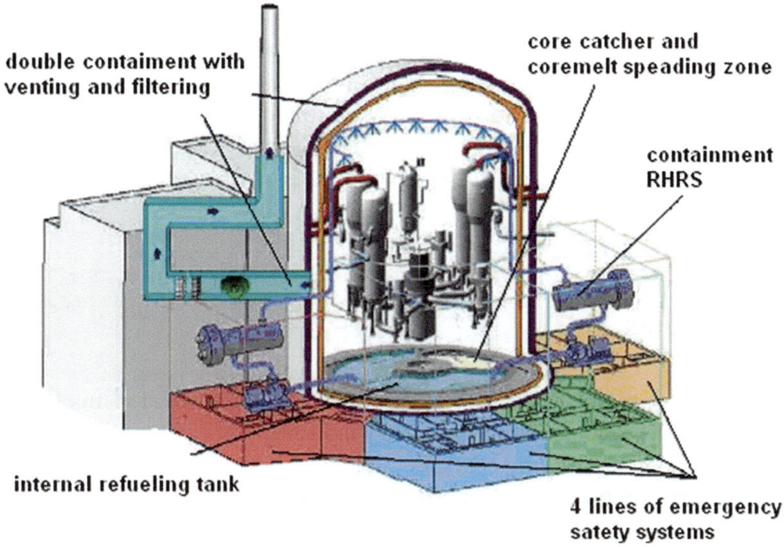

Fig. 6 EPR: GEN-III PWR NPP under construction in China

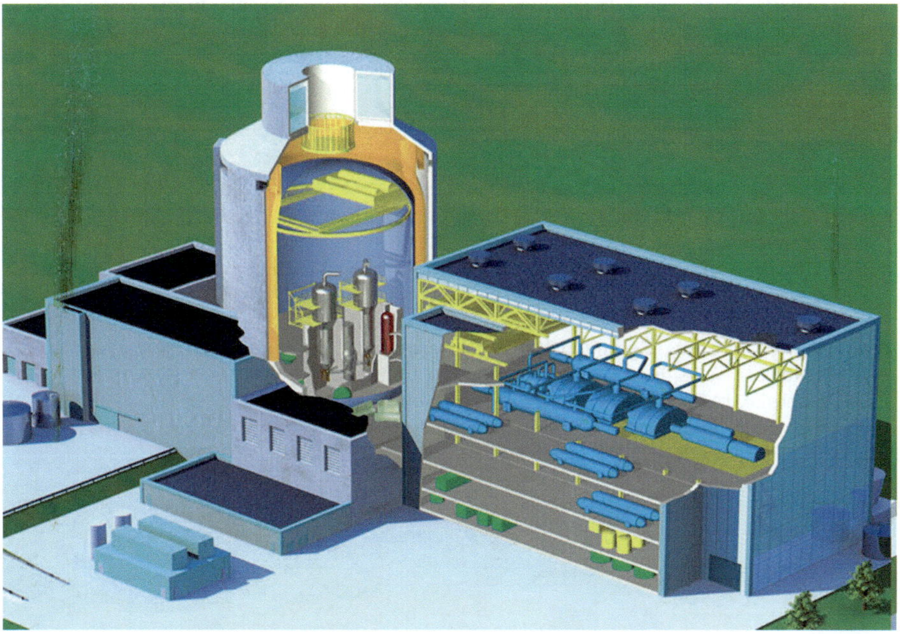

Fig. 7 AP1000: GEN-III passive safety PWR under construction in China

non-proliferation, China has enhanced its implementation of international conventions related to nuclear non-proliferation and nuclear material control. Import and export of NPP technologies are strictly controlled with license system. New and used nuclear fuels are also strictly controlled. Especially, the reprocessing, transport and storage of nuclear used fuels are also strictly controlled. IAEA safeguard for civil nuclear facilities have ensured firmly the peaceful use of nuclear technologies of China.

4 Conclusions

Nuclear safety is the most important issue for China to develop nuclear power plants and is the backbone of nuclear energy industry. The "safety first" must be put to utmost priority in expanding NPP fleet. Human being must pay sufficient respect to the objective law of the nuclear power construction based on scientific policy-making. Thus, the pace to develop nuclear power actively and stably in China in near future can return to normal track.

Acknowledgment The author were grateful for the invitation of Professor Ishihara and Dr Zhang of Kyoto University to attend the symposium and for the support of the Ministry of Education, Culture, Sports, Science and Technology of Japan via "Energy Science in the Age of Global Warming" of Global Center of Excellence (G-COE) program (J-051).

References

1. Nguyen F (2011) World energy, electricity and environment. Lecture notes on 2011 WNU one-week orientation-course, July 4–6, Beijing
2. Zhao RK, Feng YC, Xu J, Li ZN (2002) Study on sustainable strategy for developing nuclear energy in China. In: Zhao RK, Ruan KQ, Shi DH, Feng YC, Xu J (eds) Progress of the Research Works in the Energy Technology Area of the National 863 Program, Chapter 5, Section 1. Atomic Press of China, Beijing, p 309 (in Chinese)
3. World Nuclear News, First IAEA Report on Fukushima, 1 June 2011
4. IAEA website, IAEA Briefing on Fukushima Nuclear Accident, 2 June 2011
5. Wang D (2004) Future energy development and nuclear prospect of China. In: NUTHOS-6, Nara, Japan, paper PL-02, 4–8 October 2004

(ii)
Contributed Papers

An Optimization Supply Model for Crude Oil and Natural Gas in the Middle East

Hooman Farzaneh, Keiichi N. Ishihara, Nuki Agya Utama, Benjamin McLellan, and Tetsuo Tezuka

Abstract Crude oil and natural gas are major contributors to the world economy. Most of the Middle East countries are the main participants in the world energy because of their reserves, supplies and also trade markets. In this investigation, a model of optimal oil and natural gas supply has been developed for the Middle East region including main producers such as Iran, Iraq, Kuwait, Qatar, Saudi Arabia, Oman and Bahrain. To this aim, Middle East region is supposed to be organized in the form of a firm and appears in the market that oriented towards establishing an effective energy system to produce oil and gas with minimum costs subject to satisfying technical, institutional and economical constraints. The model is used to prepare a projection on oil and gas supply up to 2030. According to the results, oil production is expected to increase in the Middle East region to meet growth in consumption. It should be expected to rise by about 30 Mbbld by 2030. Also, the Middle East's share in global gas production is predicted to expand to 20% in 2030. The projection implies that the Middle East upstream and refinery capacities are likely to be sufficient to meet the demand until around 2015; thereafter expansion appears certainly.

Keywords Crude oil • Energy supply • Middle East • Natural gas • Optimization model

H. Farzaneh (✉)
Graduate School of Energy Science, Kyoto University, Kyoto 605-8501, Japan

Graduate School of the Environment and Energy, Science and Research Branch of Islamic Azad University, Tehran, Iran
e-mail: hooman.farzaneh.2v@kyoto-u.ac.jp

K.N. Ishihara • N. Agya Utama • B. McLellan • T. Tezuka
Graduate School of Energy Science, Kyoto University, Kyoto 605-8501, Japan

T. Yao (ed.), *Zero-Carbon Energy Kyoto 2012*, Green Energy and Technology, 17
DOI 10.1007/978-4-431-54264-3_2, © Springer Japan 2013

1 Introduction

Oil is the most used energy source worldwide. According to recent researches, it is clear that the world will soon face a global crisis due to dwindling oil resources and a peak in production [1, 2].

The Middle East region contains the largest quantities of proven oil and gas reserves in the world. According to OPEC, the proven crude oil reserves amount to 774.6 (billion barrels oil equivalent) bboe and proven natural gas reserves up to 505.3 bboe are also located in this region [3]. The Middle East accounts for more than 800 billion barrels of oil ultimate recoverable reserves (URR) or 42% of global URR. Saudi Arabia controls the largest share of URR, with 300 billion barrels, and Iraq and Iran follow with 135 billion barrels and 130 billion barrels, respectively. Together these larger reserve holders control more than 70% of Middle East URR [4].

Middle East countries currently contribute 14% of global energy production [5]. In former decades, the share of the Middle East in world oil production varied widely. Before the first oil crisis their share was up to 44%, in the mid-1980s it decreased to about 28% [6]. Five countries (Saudi Arabia, Iran, United Arab Emirates, Iraq and Kuwait) each produce more than 2 million barrels per day, contributing by far the largest proportion of regional production. Saudi Arabia leads in terms of oil, as do Iran, Qatar, Oman and UAE in terms of natural gas.

Increasing population, higher standards of living, accelerated growth of energy-intensive industries (cement, aluminum, petrochemicals, etc.) and highly-subsidized energy prices have led Middle Eastern demand for oil and gas to almost double [7]. The strong growth in demand for water, which is met partly by seawater desalination, also has the effect of increasing energy consumption. Also, the price of energy is politically a highly sensitive issue in this region. Some countries must now begin to face the need to balance domestic energy demand against export demand for oil and natural gas and the need to increase the price of energy to levels more compatible with new cost conditions.

It can be stated that the Middle East region has been one of the major influences on global energy markets, and the future of global energy demand and supply is strictly dependent on how the reserves are managed in this region. Having considered the outlook for supply of oil and gas from these suppliers, it becomes clear that a number of challenges lie ahead. A major one relates to the significant uncertainties over how much future production will be required and how the Middle East region could be able to balance the world oil and gas demand. Moreover, alongside this concern, there are various other challenges including the emergence of oil as a financial asset; upstream costs; the technology evolves, the issue of sustainable development and the domestic energy demand.

This paper analyzes the aforesaid concern and its consequences by estimating a cumulative supply curve for conventional oil and gas in the Middle East region through the using an optimization supply model.

2 Concept of the Model

Analysis of multi-dimensional interactions of flow of energy in the Middle East Region is a complex task that necessitates the development and utilization of analytical tools [8, 9]. Development of analytical tools with high complexity is usually based on conclusions of many concepts and theories from different scientific disciplines. The economic rationality of a producer and activity of a firm has extensively been developed as a branch of microeconomics and it has been utilized for explaining the development of production technology. In this case, the Middle East region can be supposed to be organized in the form of a firm and appears in the market that is oriented towards establishing an effective energy system which may be identified as producing oil and gas with minimum total cost subject to satisfying the demand and other technical, economical and institutional constraints. This concept may be formulated as below [10]:

$$Min\ TC = \sum_i p_i F_i \tag{1}$$

Subject to:

$$f(F_1, \ldots, F_n) \geq Q \tag{2}$$

$$\sum_i F_i \leq A_i \tag{3}$$

$$F_i \geq 0 \tag{4}$$

where, TC, p_i, F_i, Q and A_i are defined sequentially as total system cost, the unit price of production factor i, production factor i, energy demand and available resources.

Application of the model is subject to the identification and estimation of the production function (Equation 2). Difficulty of obtaining an explicit form of the production function has in many cases led to an indirect solution of the model. This is usually achieved by defining total production function as a function of the parameters of the model. The production function represents a nonlinear system and it is a complex function. Empirical estimation of the complex function would be associated with considerable over or underestimations. It is, therefore, possible to segregate the Middle East energy system into incremental elements and substitute the non-linear production function with a set of interrelated equations based on the theory of system integration. Simultaneous solution of the set of equations would help identifying the function of the whole system and state of variables at different points of the system and time can be estimated. Such a formulation would enable to solve the behavioral model of the production firm explicitly. Implementation of system integration in the Middle East region may be represented by the help of the reference energy system (RES) diagram as depicted by Fig. 1.

Level of segregation is usually determined by the ability to introduce the oil and gas producing process or different processes of energy conversion. Therefore, a set

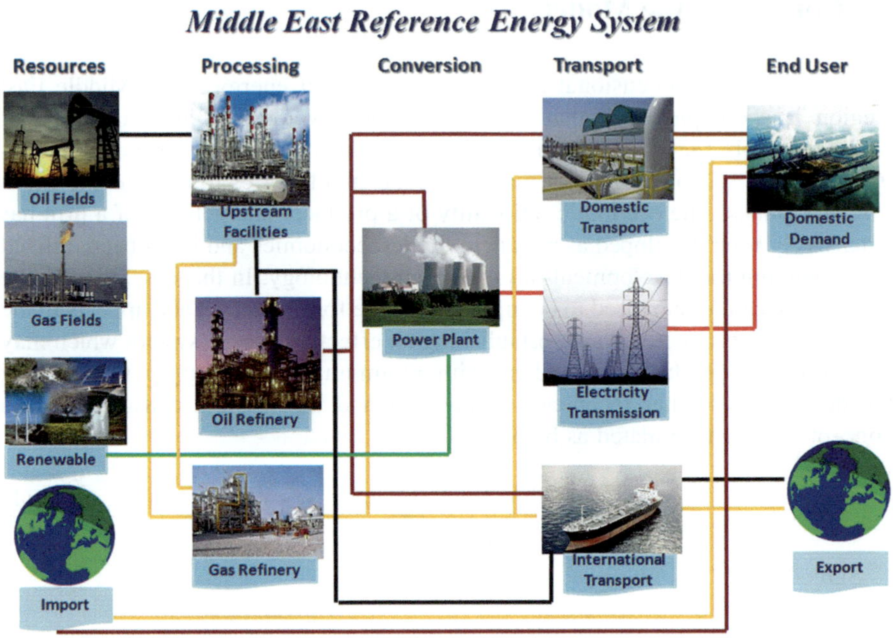

Fig. 1 Middle East region RES

of simultaneous equations would be developed that represent the laws governing the flow of energy through different sub-systems (control volumes) and levels at different operating conditions in the Middle East region. Also, defined technologies at different levels represent the functionality of each sub-system in the whole system.

2.1 *Constraints of the Model*

According to the Middle East RES, the following description of the energy flow constraints is given for the set of level identifiers as: U (useful energy), F (final energy after distribution), T (final energy after transmission), X (secondary energy), *A* (primary energy) and *R* (energy resources). Some of the important constraints of the model can represent through the following equations:

A) Demand constraint:
 Demand must be met. In each region (countries) the supply of crude oil, oil products and natural gas must be greater than or equal to the demand of those fuels in that region:

$$\sum_{fvlt} \eta_{fvlt} U_{fvlt} \geq U^*{}_{fvlt} \tag{5}$$

where, η, U^* represent efficiency and defined energy demand for energy carriers f through using technology v in each region during the load region l and time period t.

B) Availability of resources:
More resources cannot be extracted from a region than exist in that region. For each region, fuel, load and time period:

$$\sum_{rglt} R_{rglt} - \sum_{fvlt} A_{fvlt} \geq 0 \tag{6}$$

where, g is the cost grade of the energy resource.

C) Speed of resource extraction:
The percentage of available resource extracted in a given year cannot exceed a defined fraction of the remaining resources:

$$A_{fvlt} \leq \alpha \left(URR - \sum_{\tau=1}^{t-1} A_{fvl\tau} \right) \tag{7}$$

α is the percentage of remaining resource that can be extracted in a year. This constraint governs the overall shape of increasing and decreasing production of a resource, representing a peak oil curve of resource depletion.

D) Capacity constraint on production:
The production of energy carriers in each region must be less than or equal to the production capacity:

$$\frac{A_{fvlt}\varepsilon_{fvlt}}{\Delta l} \leq \sum_{\tau=1}^{t} H_{fvl\tau} PF_{fvl\tau} + \sum_{\omega=b-((PL-t))}^{b} H_{fvl\omega} PF_{fvlt} \tag{8}$$

where PF and PL are plant factor and plant life of each sub-system. H is the production capacity.

2.2 *Objective Function of the Model*

Costs represent the extent of allocation of factors and resources to the Middle East region. A criterion that is used for identifying an optimal point of development is minimum usage of total factors and resources. Such a criterion is the dual of economic efficiency of the Middle East region which is the maximization of profits. The cost function is the sum of the present values of capital, maintenance, operation, resources, and externalities (social costs) of the whole Middle East energy system which may be represented by the following formula:

$$TC = \sum_{t} \beta_m^t \Delta t \left\{ \sum_{v} \sum_{l} \sum_{f} \left(E_{fvlt} \varepsilon_{fvl} CV_{fvlt}\right) + \left(H_{ft} CF_{fvlt}\right) + \sum_{\tau=t-\tau_{flvt}}^{t} \left(H_{f\tau} CF_{fvl\tau}\right) + \sum_{r} \left[\sum_{g} \sum_{l} R_{rglt} \left(c_{rglt} + x_{rglt}\right) + \sum_{c} \sum_{l} \left(I_{rclt} ci_{rclt}\right) - \sum_{c} \sum_{l} \left(EX_{rclt} ce_{rclt}\right) \right] \right\} \tag{9}$$

β_m^t is the discount factor in the period and is calculated as:

$$\beta_m^t = \Pi_{i=1}^{t=1} \left[\frac{1}{1 + \frac{dr(i)}{100}} \right]^{\Delta i} \tag{10}$$

where, $dr(i)$ is the discount rate. c_{rglt}, ci_{rclt}, ce_{rclt} and x_{rglt} indicate the unit price of: resource, import, export and externalities such as carbon tax in each region. E represents the energy flow in different levels. EX is the energy export flow.

2.3 *Concept of Type Curve of Oil and Gas Fields*

In the Middle East region, utilization of a field of crude oil or natural gas follows the physical characteristics of the field. According to Equation (7) production varies as a function of accumulated extraction of oil and gas fields. The pattern of production of rich gas from a field may be described according to the Hubbert peak oil theory as follows [11]:

$$R = \sum_{i=1}^{k} (R)_i = \sum_{i=1}^{k} 4(R_{\max})_i \left\{ \frac{e^{-\alpha(t-t_{\max})}}{\left[1 + e^{-\alpha(t-t_{\max})}\right]^2} \right\}_i \tag{11}$$

And:

$$\alpha = 4\frac{R_{\max}}{URR} \tag{12}$$

where in the above equations, k, $t_{\max}$ and $R_{\max}$ are defined sequentially as number of defined cycles for peak estimation, peak time and peak production of the resource.

3 Results and Discussions

The aforesaid model has been developed to show how the future might look on the basis of perpetuation with current energy policy in the Middle East region. Therefore, the model is expected to use historical trends of demographic and energy related data in the Middle East countries to project the energy (crude oil and natural gas) supply up to 2030.

3.1 Assumption

Population growth is the key driver of the future energy trends as the level of the population has a direct effect on the size of energy demand and an indirect effect by influencing economic growth and development. The region's population has grown from 83 million in 1970 to 191 million in 2010. Projections from the United Nations shows an average per annual population growing by around 1.9% from 2010 to 2030 [12]. Also, projections from the International Monetary Fund (IMF) shows global gross domestic product growing by around 4.1% in 2010. The GDP is projected to grow an average per annual by 3.5% from 2010 to 2030 [13]. The evolution of energy prices is a key determinant of energy export and trade in the Middle East region. In this paper, the assumed energy price is considered on the basis of IEA energy outlook 2011. Oil price is assumed to rise steadily to 2030 as rising global energy demand requires the development of increasingly expensive sources of crude oil. Natural gas price has been closely correlated to the oil price in long term oil supply. The natural gas price is assumed on the basis of import prices in different regions such as US, Europe and Japan.

Energy intensity is assumed to increase 1.5% annually from 2010 to 2030 in the Middle East region. There are regional differences in energy intensity in this region. Iran, Iraq and Saudi Arabia are expected to increase their energy intensity while UAE shows a steady growth in its energy intensity [14].

3.2 Oil and Natural Gas Maximum Production

In this paper, historical trends of oil and gas production in eight main producers in the Middle East region has been used as input data. Also, URR is estimated at about 833 billion barrels crude oil according to study technical reservoir reports of 304 active rigs and 1,031 completed wells in this region [15]. According to the model results, the maximum oil production from the Middle East region is estimated 32 mbd by 2019 and it will be reduced to 30 mbd by 2030. The peak year of natural gas production is estimated by around 2048 with a maximum rate about 2.24 tcm. Comparison between oil and gas production is shown in Fig. 2.

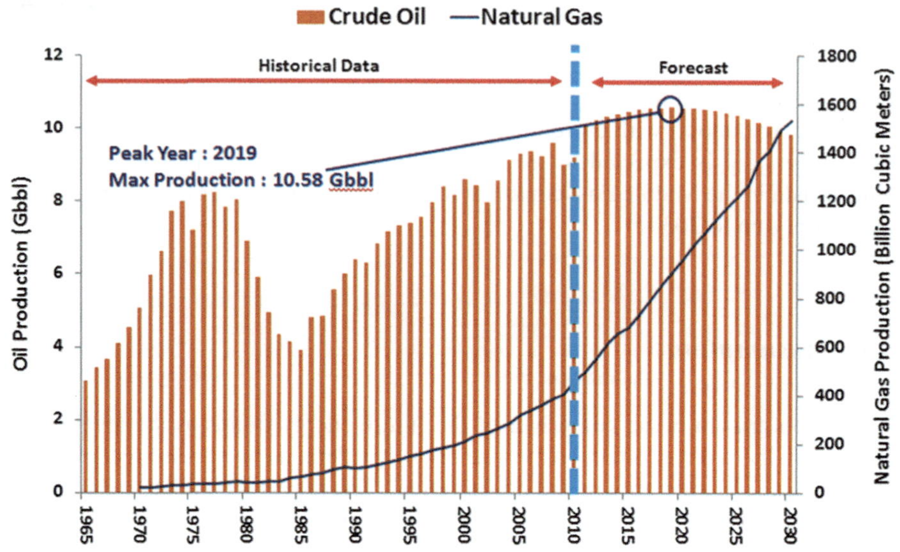

Fig. 2 Oil and gas maximum production in the Middle East region

3.3 *Supply–Demand Match*

The primary energy supply is estimated on the basis of the projected energy demand by OPEC in the Middle East region [15].

According to OPEC, energy demand is increased by around 96% from 2010 to 2030. The OPEC projection in which only existing policies are assumed to be in effect, sees oil demand reaching 502.3 Mtoe by 2030, a 40% increase over the 2010 levels or an average annual increase of 1.9%. Demand for natural gas increases 140% from 360.2 Mtoe to 502.3 Mtoe between 2010 and 2030. This increase in demand for natural gas is primarily due to the gradual shift in the power plants away from oil products to natural gas. Consequently, energy supply grows steadily through the projection period, from 1,593 Mtoe in 2010 to 2,202 Mtoe in 2030, a 38% increase over the 2010 levels as depicted by Fig. 3.

The supply–demand match shows lower increase in primary energy supply compared with energy demand in the Middle East region. Therefore, oil and gas supply will not be sufficient to support export in long term projection (Fig. 4).

Iraq's oil production potential is immense, but exploiting it depends on consolidating the progress made instability. Iraq remains one of the least explored major resource holders. Although about 80 fields have been discovered, only 35 are currently in production.

Saudi Arabia is expected to contribute the second largest increase in oil production in the Middle East region, raising production by almost 33% from 10.1 Mbbld in 2010 to nearly 13.4 Mbbld by 2030 as shown in Fig. 5.

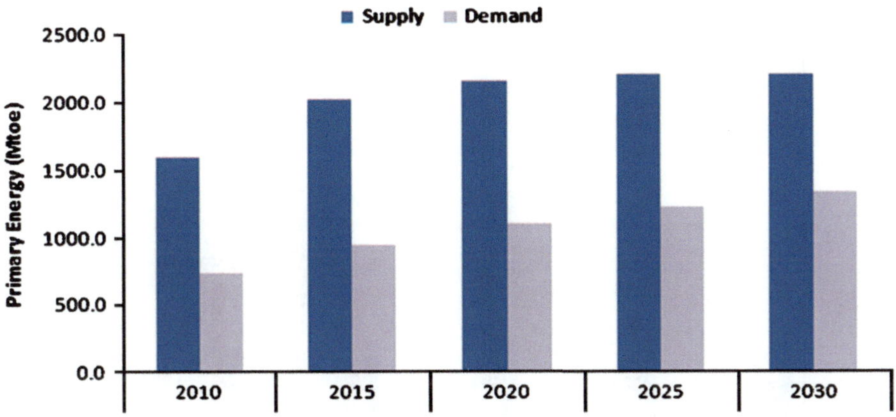

Fig. 3 Primary energy supply–demand match in the Middle East region

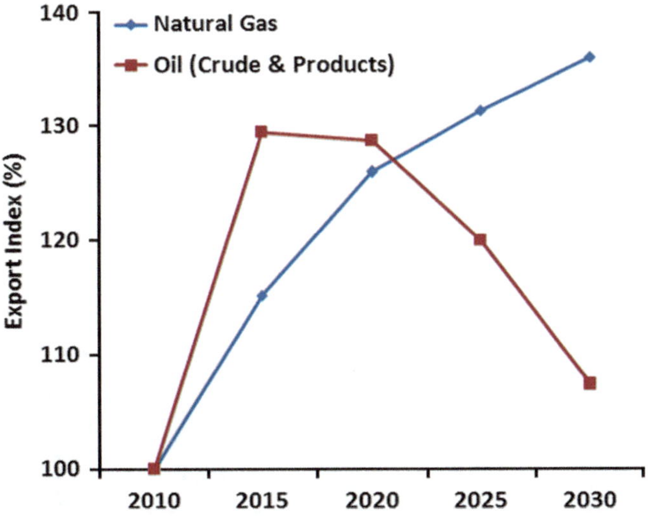

Fig. 4 Natural gas and oil export by the Middle East region up to 2030

The region's share in global natural gas production is expected to expand form 14% in 2010 to 20% by 2030 as depicted by Fig. 6. Policies promoting fuel diversification support a major expansion of global gas use. This is met through higher domestic production and through an increasing share of LNG trade and pipeline import from the Middle East region. In this case, Qatar, Oman, UAE and Iran remain the largest gas producers in 2030.

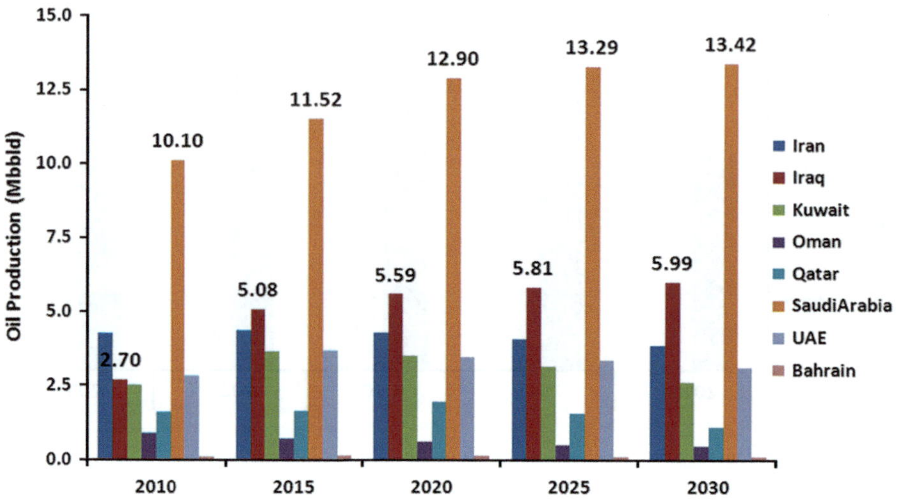

Fig. 5 Oil supply by countries up to 2030

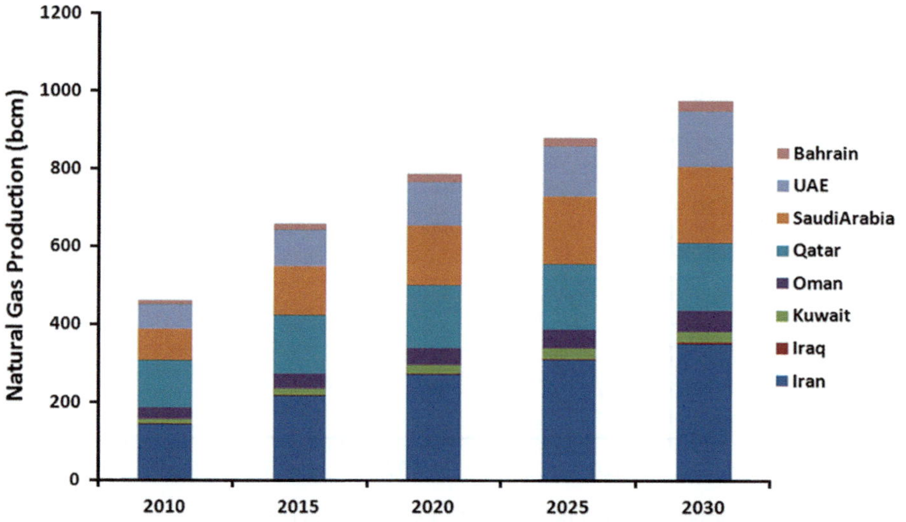

Fig. 6 Natural gas supply by countries up to 2030

3.4 Additional Capacity and Investment Requirement up to 2030

Middle East region's oil resources remain the cheapest to exploit, although producing investments are increasing in this region. The model results show that the cumulative investment in the oil upstream is expected about 755 billion dollars up to 2030.

Table 1 Energy supply and cumulative investment in the Middle East up to 2030

	Supply		Cumulative investment (billion $)
	2010	2030	2010–2030
Crude oil (Mbbld)	25.5	29.7	755
Oil product (Mbbld)	7.1	12.5	120
Natural gas (bcm)	462.3	976.6	267
Total			1,142

Oil refinery capacities must increase to meet the domestic demand and export. Results show that the highest additional capacity in the oil refinery industry in the middle East region happens during 2010–2015 which is largely a result of increasing oil production to reach the peak point by 2019. The cumulative investment during 2010–2030 is estimated about 120 billion dollars.

Projected trends in gas demand and supply call for total cumulative investment around 267 billion dollars up in supply infrastructure and also trade by LNG and pipeline. Summary of energy supply and required cumulative investment is represented in Table 1.

3.5 Restricted Investment Scenario

According to the current policy (investment scenario), the Middle East region has been considered to contribute all of the required growth in oil production up to 2030. To achieve this growth, upstream investment in the Middle East needs to average $38 billion per year from 2010 to 2030. But it is far from certain that all of this required investment will be forthcoming for many reasons affecting some of the countries in the region such as: political instability, infrastructure, national financing constraints and economic sanction.

In this case, a scenario has been defined through developing the model to analyze how the global oil market might evolve if a restricted Middle East upstream investment occurred. Figure 7 shows the expected results.

As the above figure shows, the fall in oil production compared with the current policy reaches over 6.9 Mbbld in 2020 large enough to have significant impact on global oil balance, increasing energy prices.

4 Conclusion

Future development in the Middle East oil and gas supply system will be affected by the rate of growth in energy demand and some challenges and uncertainties about investment requirement. Limits on rates of capacity expansion will strongly

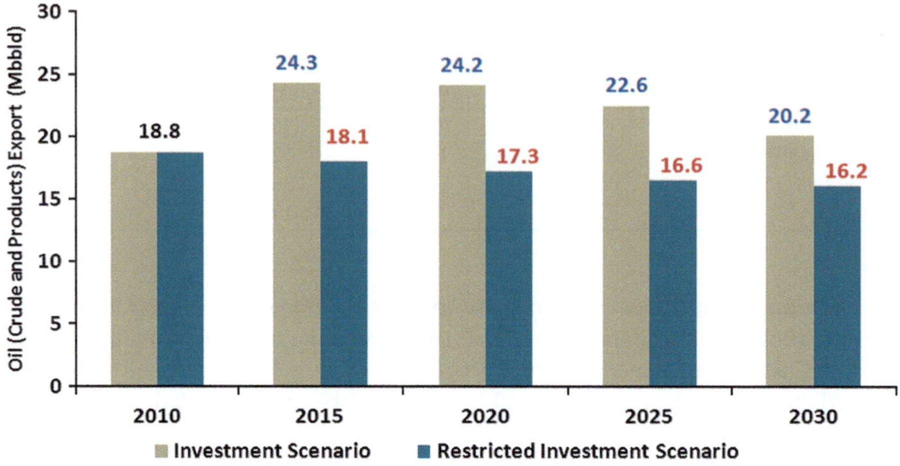

Fig. 7 Restricted upstream investment scenario

affect the fuels which should be produced by this region in the future. Towards the end of 2030, the Middle East should increase its role as an exporter of hydrocarbons if Iraq and Saudi Arabia are able to exploit their enormous reserves effectively. Iraq crude output is projected to grow from about 2.7 Mbbld currently to more than 6 Mbbld. Saudi output is likely to expand by nearly 3.2 Mbbld.

Middle East gas production share of global production is expected to expand by 6% through the next 20 years if Iran and Qatar the region's leading players in natural gas decide to consolidate their positions.

Capital investment will be required for additional capacities in the oil and gas industry. Restricted Middle East oil investment may decrease oil export by 16% and also increase risk of oil product imports by more than 40% in future.

Acknowledgment The authors would like to thank the GCOE program in Graduate School of Energy Science, Kyoto University for the great support and helpful comments.

References

1. Harrigan J, Chengang W, El-Said H (2006) The economic and political determinants of IMF and World Bank lending in the Middle East and North Africa. World Dev 34(2):247–270
2. Hinnebusch R (2003) The international politics of the Middle East. Manchester University Press, Manchester
3. Organization of the Petroleum Exporting Countries (2010) World Oil Outlook. ISBN: 978-3-9502722-1-5
4. Campbell CJ (2002) ASPO database for 64 major oil-producing countries, UK-based Oil Depletion Analysis Center
5. BP Statistical Review of World Energy (2011) www.bp.com/statisticalreview
6. Middle East Economic Survey (2012) http://archives.mees.com/issues/59/articles/2192

7. International Energy Agency (2011) World Energy Outlook. ISBN: 978 92 64 12413 4
8. Kian B, Pourfakhari MA (2010) A system dynamic model for production and consumption policy in Iran oil and gas sector. Energy Policy 38:7764–7774
9. Brandt A, Plevin R, Farrel A (2010) Dynamics of the oil transition: modeling capacity, depletion and emission. Energy 35:2852–2860
10. Farzaneh H, Dashti M (2010) Integrated modeling framework for optimizing energy demand in high energy intensive industries of Iran. In: IEEE International Energy Conference. ISBN: 978-1-4244-9380-7/10
11. Nashavi I, Malallah A, Al-Bashir M (2010) Forecasting world crude oil production using multicyclic Hubbert model. Energy Fuel 24:1788–1800
12. United Nations department of economic and social affairs (2010) http://esa.un.org/unpd/wpp/p_wpp/htm/pwpp_Total-Population.htm
13. International Monetary Fund (2011) http://www.imf.org/external/pubs/ft/weo/2010/02/map/?year=20108region=1458datacoll=PPPSH
14. World Bank (2010) World development indicators. http://data.worldbank.org/indicator/EG.USE.PCAP.KG.OE
15. Organization of the Petroleum Exporting Countries (2011) Annual statistical bulletin. ISSN: 0475-0608

Energy Demand Forecast for South East Asia Region: An Econometric Approach with Relation to the Energy Per Capita "Curve"

Nuki Agya Utama, Keiichi N. Ishihara, Tetsuo Tezuka, Hooman Farzaneh, Benjamin McLellan, and Qi Zhang

Abstract This paper will discuss the forecast of final energy consumption from each member country up to 2100. Almost 40 years (1971–2010) of data series of energy consumption had been collected from various sources. Econometrics was then used to predict the energy consumption, based on the population projections of the United Nations (UN) (medium scenario). An energy intensity index (KTOe per capita) from various developed countries was also used (as a reference) in order to measure the potential limits of ASEAN member countries' energy consumption. A novel approach was taken to modifying standard econometrics approaches with further limiting factors also being defined, such as land to population ratio, population trend, total land area, demographic characteristics and landscape (inland, archipelago, landlocked etc.).

It is predicted that in 2100 the region will require 3.5 GTOe in the medium population scenarios respectively.

Keywords ASEAN • Demographic • Econometrics • Energy demand • Energy per capita

1 Introduction

The Association of South East Asia Nations (ASEAN) consists of ten member countries—namely, Brunei Darussalam, Cambodia, Indonesia, Lao PDR, Malaysia, Myanmar, Philippines, Singapore, Thailand and Vietnam—and covers a total land area of 4.34 million square kilometers. It had a total population of 566 million in

N.A. Utama (✉) • K.N. Ishihara • T. Tezuka • H. Farzaneh • B. McLellan • Q. Zhang
GCOE Energy Scenario Planning Office, Graduate School Energy Science, Yoshida-Honmachi-Sakyo ku, Kyoto University, Kyoto, Japan
e-mail: agyautama@yahoo.co.uk

T. Yao (ed.), *Zero-Carbon Energy Kyoto 2012*, Green Energy and Technology,
DOI 10.1007/978-4-431-54264-3_3,

2007 (8.6% of total world population) and is predicted to have more than 1.4 billion residents in 2100.

The energy consumption has increased more than fivefold, from less than 100 MTOe to more than 562 MTOe in 2010. The region's final energy consumption has been increasing at an annual rate of 3.8%. The industrial sector has had the fastest growth at an average annual rate of 6.1%, as cheap labor and political stability have been the major driving factors for investment. Meanwhile, the transportation sector accounted for the second largest growth (3.5% on average) mainly due to population growth and economic development. However the growth had decreased as a share of the total from 27.7% in 1995 to 23.9% in 2007. The remaining residential, commercial, transport and non-energy sectors had a growth rate of 2.2%. Furthermore, looking to the future, the oil consumption is predicted to increase threefold from 185 MTOe in 2007 to 496 MTOe by 2030 [1].

The reserves of natural gas, for instance in Indonesia and Malaysia alone, are proven to be more than 5.5 TCM (tera cubic meters) or around 30% of the reserves available (over 15 TCM) in the whole Asian region. The total reserves of more than 4,300 MMT (million metric tonnes) coal in Indonesia and Thailand (both bituminous and lignite) represent the biggest fossil fuel reserves in the region [2]. However, these reserves are relatively low compared to the worldwide reserves or, for example, China which holds the third largest coal reserve (ca. 115,000 MMT). The region's total final coal consumption increased from 248.7 MTOe in 1997 to 1,620 MTOe in 2006, to meet the electricity needs which have rapidly increased from 369 TWh in 2000 to 3,600 TWh in 2010 [2].

The oil price boom in 2007–2008 was the crucial moment for policy makers in ASEAN member countries to consider reducing the region's dependence on fossil fuels by shifting to renewable energy resources. According to the prediction by the Asian Center of Energy (ACE) [2], the share of generation mix in the region will move towards non-oil fuels. But by 2020, almost 45% of the fuel for power generation in ASEAN will still be coal, followed by 40% natural gas and only less than 2.0% oil. The rest of the electric power will be generated either by renewable energy or nuclear power [3].

Moreover, a least cost optimization model developed for Indonesia, Malaysia and the Philippines [4] shows that the room for renewable energy as part of the power generation mix is very promising compared to nuclear power in 2030, however from 2050 onwards nuclear power will be cost effective enough to compensate for the gap in fuel supply from depleted fossil fuel reserves.

Based on the causality analysis completed for the ASEAN region, macroeconomic factors have a strong relation with increasing the power demand. The bi-directional relationship from energy causing the increase of economic growth and vice versa has been empirically proven [5]. Moreover, apart from the increase of GDP influencing the energy consumption, the increase in population also significantly influences the increase in energy demand through the transportation sector [6], power demand through urbanization [7] and through the expansion of the residential sector [8]. These economic and population factors are very crucial variables in predicting the future energy consumption in this vast developing region (Table 1).

It is most likely with current economic growth and development the ASEAN member countries energy per capita will remain steady when reach its saturation, similar condition shows by developed countries as seen in Fig. 1.

Most developed countries such as United States and Sweden has reached its steady energy per capita basis since more than 30 years ago, while Japan and Italy reached its maturity in average of 4.1 and 2.7, respectively, in the last 15 years. These figures can be a benchmark for predicting the future developing countries in ASEAN member countries. However some measure should also be considered such as similarity of the landscape, population and population distribution, transportation between populations or between region and scale of the countries (as seen in Table 2).

For instance the land scale of the US (9.8 million km^2) compare to Indonesia (1.9 million km^2), if we includes the sea coverage the size of Indonesia will reached up to 7.9 million km^2 almost close to the overall size of the US area. This condition may lead in the future to high consumption of energy through sea/water transportation. While in the case of Philippines, Malaysia, Thailand, Vietnam and Myanmar which have land size of 300,000, 330,000, 513,000, 331,000 and 676,000 km^2, respectively, these conditions are relatively similar to the countries such as Japan, Sweden and France (between 377,000 up to 676,000 km^2). Moreover the landscape conditions for these countries are also similar, where the coastline dominates the countries' borders rather than land borders with neighboring countries. The conditions are rather different in the case of Lao PDR and Cambodia, where the size of these two countries are less than 250,000 km^2. However Lao and Cambodia cannot be compared in terms of size and have significant differences in terms of their borders—Lao PDR are a landlocked country while Cambodia has coastlines.

The studies shows the predicted energy demand in ASEAN member countries by using demand model and unconventional parameters such as landscape, geographical boundaries and decentralized/centralized population distribution apart from GDP and population growth (Fig. 2).

2 Methodology

The boundaries between archipelagos, landlocked and small-large islands should be defined clearly in the beginning, for instance in case of wide archipelagos (e.g. Indonesia) they require both land and sea transportation, therefore there are some major challenges to develop infrastructure (mainly for sea infrastructure). The infrastructure is not only limited to developing seaports, but also supporting facilities (storage, docks, access routes etc.) and bridges (small to large). This type of country requires high-energy demand since using large-scale sea-freight or smaller ships will transport most of the goods. The lengths of bridges will also increase the energy consumed. For landlocked countries such as Lao PDR, land

Table 1 ASEAN demographic, social, economy and energy indicators (2010)

	GDP (billion US$)			Population (thousand)			Energy consumption (KTOe/capita)			
Country	1990	2010	Growth (20y)	1990	2010	Growth (20y)	1990	2010	Growth (20y)	Total area (thousand sq km)
Brunei Darussalam	3.5	14	300%*	252	399	58%*	6.99	7.50	7%	5,765
Cambodia	2.2	11.24	411%*	9,532	14,138	48%*	n/a	0.32	n/a	181,035
Indonesia	114.4	706.6	518%*	184,346	239,871	30%	0.55	0.73	33%	1,904,569
Lao PDR	0.865	7.29	743%*	4,192	6,201	48%*	n/a	n/a	n/a	236,800
Malaysia	44	237.8	440%*	18,209	28,401	56%*	1.21	2.02	67%*	329,847
Myanmar	2	19	850%*	39,268	47,963	22%	0.27	0.28	2%	676,578
Philippines	44.3	119.6	170%	61,629	93,261	51%*	0.47	0.52	12%	300,000
Singapore	36.1	208.7	478%*	3,017	5,086	69%*	3.80	4.91	29%	683
Thailand	85.3	318.5	273%*	57,072	69,122	21%	0.73	1.15	56%*	513,115
Vietnam	6.5	106.4	1537%*	743	1,124	51%*	0.36	0.47	29%	331,689
World	21,900	63,120	188%	5,306,425	6,895,889	30%	2.27	3.12	37%	n/a

*Growth rate above world average

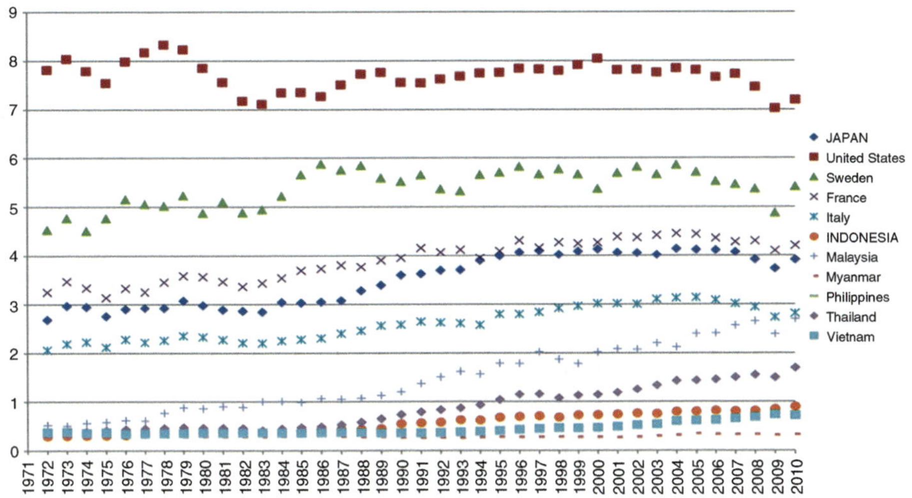

Fig. 1 Comparison between developed and ASEAN member countries energy per capita (TOe/capita)

transportation will be the only option, therefore the energy consumption from this sector is relatively easy to predict and relatively smaller than island or archipelago countries. The following

$$R_i = cE_n \gamma g \tag{1}$$

cE_n is the ratio between energy intensity (EI) of the referenced country (developed countries) and the EI from the selected country (ASEAN). The energy per capita (Ep) from the reference country will be selected if more than 80% of the ratios are in the range $-10\% \leq \text{ratio} \leq 10\%$ as seen in Fig. 3 (left with red dot) the ratio of EI between the US (as reference's country) and Indonesia (as selected country), more than 80% of the ratio are within the range of $\pm 10\%$, therefore the information of Ep cannot be used for predicting Indonesian Ep. In contrary as seen in the same figure (right with blue dotted) between Japan and Indonesia shows that more than 20% of the ratios are beyond the range of $\pm 10\%$, therefore Japan can be a benchmark for predicting the future Ep for Indonesia. Furthermore if we found more than one case of ratio in between $\pm 10\%$, the Ep from the multi-references was accounted as an average.

Figure 3 shows a population density (people/square meter) and total population of the referenced and selected countries, where category II and III are located in the similar group, while the detail conditions are assuming to be different (as explained above), while category I is located rather separated form other countries. These shows that most likely the future selected ASEAN energy per capita will follow the current energy per capita of the referenced countries.

Table 2 Parameters to be considered other than energy per capita

	Scale	Population	Landscape	Reference	ASEAN	Note
I	Large	Large	Long coastal lines	US, Canada	Indonesia	High energy per capita; from transportation and population
II	Medium	Medium	Inland with long coastal areas	Italy, France, Sweden, Japan,Korea	Malaysia, Thailand, Philippines, Vietnam, Myanmar	Medium energy per capita; from transportation and population
III	Small-medium	Small to medium	Small coastal line, no coast or deadlock	Switzerland, the Netherlands	Lao PDR, Cambodia, Singapore and Brunei	Low energy per capita; from transportation and population

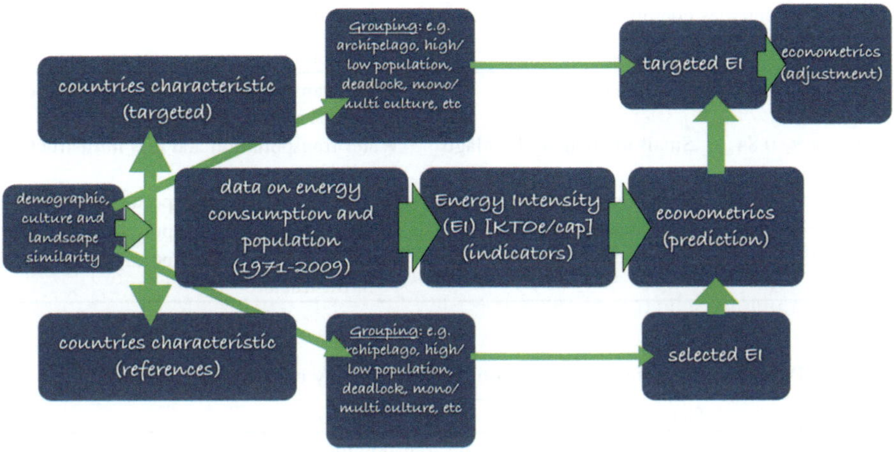

Fig. 2 Methodology flow chart for forecasting the final energy consumption in the region

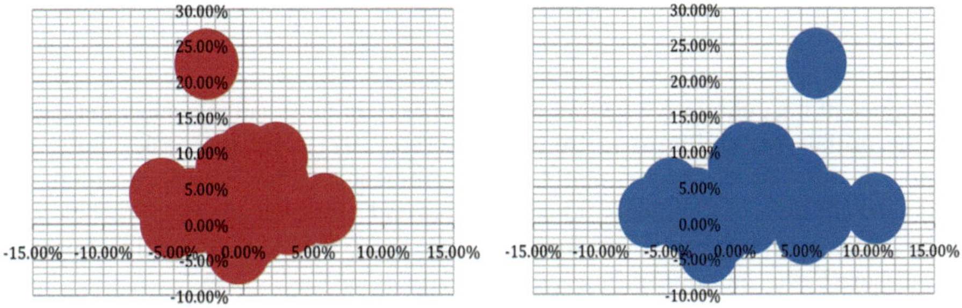

Fig. 3 The energy intensity ratio between referenced and selected countries (an example)

To calculate R_i, we start with the data from the energy intensity growth between selected countries compared with the referenced countries. The cumulative amount of energy per capita should refer to the reference countries (developed countries) selected based on the demographic and geographic conditions, whose indices are explained in Tables 3 and 4.

$$E_o = \left(\frac{R_i}{E_i}\right) - 1 \tag{2}$$

$$r = \left(\frac{\alpha}{6} R_i\right) e \tag{3}$$

$$t_e = t_o \left(\frac{1}{\alpha}\right) \ln(E_o) \tag{4}$$

Table 3 Boundaries condition for type of country by landscape and geographical conditions

Boundaries	Landscape (dominated by)	Characteristic
$0.85 \leq g \leq 1.00$	Wide archipelago	Water transportation and relatively large infrastructure challenges
$0.75 \leq g \leq 0.84$	Small-medium archipelago	Water transportation and less infrastructure challenges
$0.50 \leq g \leq 0.74$	Small medium land	Dominated by land transportation
$0.25 \leq g \leq 0.49$	Landlocked land	Dominated by land transportation
$g \leq 0.24$	Small island	Dominated by land transportation and good infrastructure

Table 4 Boundaries condition on the relation between energy intensity (EI) and GDP

Boundaries	Current EI	Current GDP per capita
$0.00 \leq \gamma \leq 0.25$	$EI \leq 0.30$	Water transportation and relatively large infrastructure challenges
$0.25 \leq \gamma \leq 0.50$	$0.31 \leq EI \leq 0.50$	Water transportation and less infrastructure challenges
$0.50 \leq \gamma \leq 0.75$	$0.51 \leq EI \leq 1.00$	Dominated by land transportation
$\gamma \geq 0.75$	$EI \geq 1.01$	Dominated by land transportation

Where:
E_o represent the cumulative factors of energy per capita; E_i is the energy per capita on the initial year (initial data from time series, 1971); r is the peak energy per capita in the predicted year, and e is the efficiency factor which considers the following issues: technology advancement, end use management (efficiency) represented by $0.1 \leq e \leq 1$ (the lower the value of e, the more efficient technology is measure being introduced in the future) (Fig. 4).

t_e represent the predicted year; t_o is the selected year or base load data (2010) and R_i is the reference countries' cumulative energy consumption. While α is the r^2 from the linier function between production (y axis) and production per cumulative production (%) in x axis.

3 Result and Discussion

Due to similarity of the open-market type of economical backbone and the global modern life style, the increases of energy per capita for developing ASEAN member countries is predicted to have a high possibility of becoming similar to current developed countries. The results of the modeling of the future energy per capita for developing countries in ASEAN with the consideration to finding the "peak" and steady growth is shown in Fig. 5. Singapore is categorized as a developed country and therefore the growth will not significantly change for the next 90 years. Malaysia, Thailand and Vietnam are predicted to reach the peak around 2065–2071, while Lao PDR, Cambodia, Myanmar and Indonesia will still

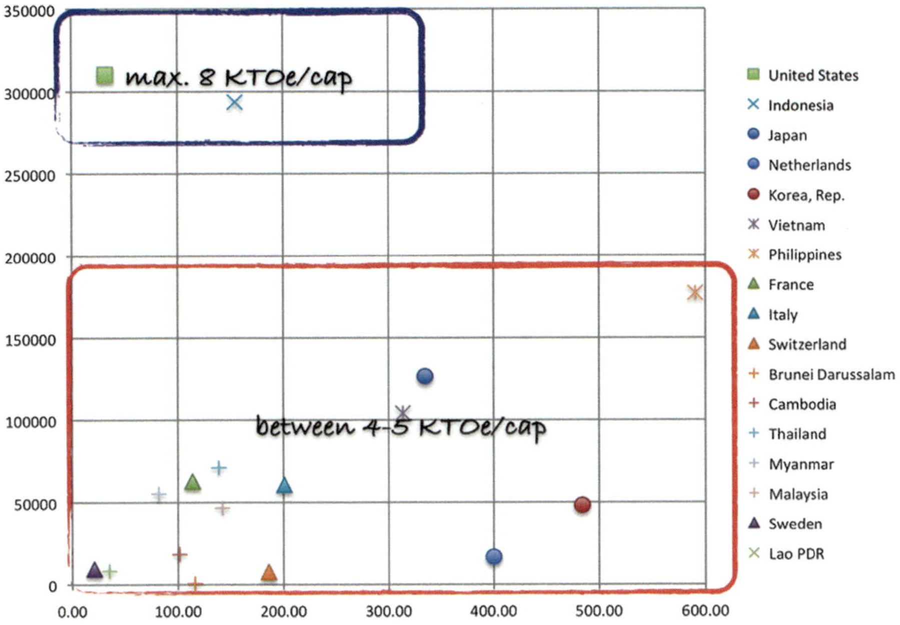

Fig. 4 Input information from reference countries based on energy capita, population and population density

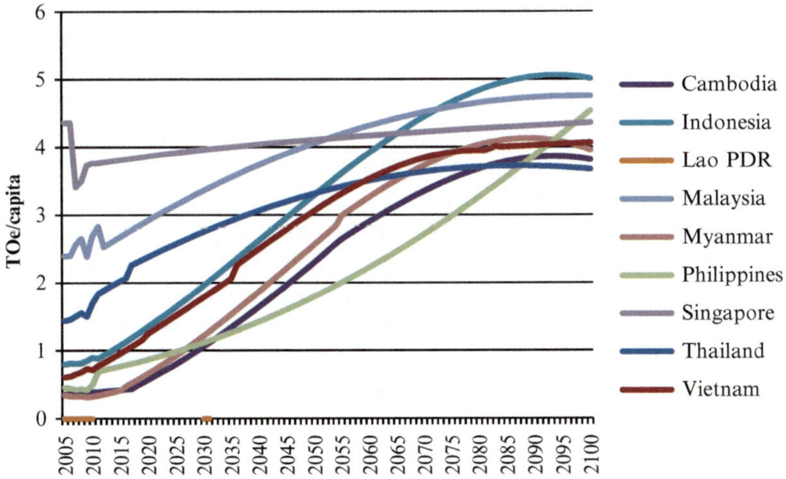

Fig. 5 Predicted energy per capita for ASEAN member countries

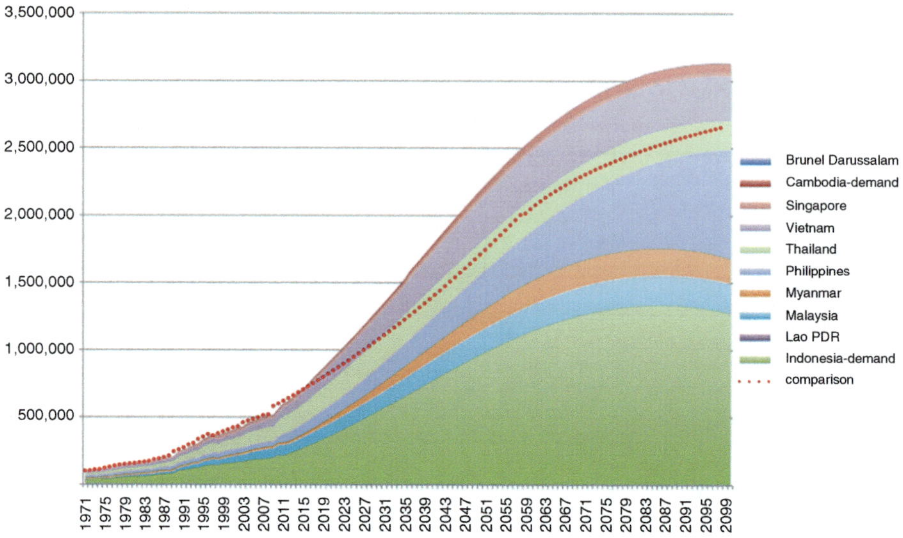

Fig. 6 Predicted final energy demand for ASEAN member countries

have significant energy per capita growth up to 2080 and remain steady afterwards. The Philippines shows an unusually strong growth pattern, with the energy per capita continuing to grow up to 2100 and beyond. This long-lived growth is influenced by the slow growth during the initial 40–50 years.

While most BaU (business as usual) or baseline scenarios predict the energy demand only based on the economic and population variables. This model has also considered landscape, geographic conditions and the size of the land-sea areas. We found that the prediction for ASEAN increased more than 17% compared to the baseline scenarios (dotted red in Fig. 6). The landscape conditions (coastal line) and size of the land-sea areas have a significant contribution to the differences. The introduction of other variables apart from economic and population growth proved to give a different prediction for the future ASEAN member countries' energy demands.

Moreover, these figures are important in order to predict the gap between productions of traditional primary energy, which shows that the countries will exhaust their current dominant primary energy sources (fossil fuels) in the next 40 years [9]. As countries which rely heavily on imported fossil fuels, the ASEAN member countries should consider in advance how much they will need to import fossil fuel in the future and what kind of alternative energy sources can be developed within the region.

Predicting the future energy demand in ASEAN by considering additional parameters such as geographical and landscape challenges while referring to the developing countries shows a significant differences, the differences caused by energy requirement for countries with demographical conditions differ with the references, such as Indonesia and Philippines, where multiple transportation mode is required such as air, land and sea transport.

Acknowledgment The authors were grateful for the support of the Ministry of Education, Culture, Sports, Science and Technology of Japan via "Energy Science in the Age of Global Warming" of Global Center of Excellence (G-COE) program (J-051).

References

1. The Institute of Energy Economic Japan (IEEJ) and ASEAN Center of Energy (ACE), the 3rd ASEAN energy outlook, February 2011
2. ASEAN Center for Energy (ACE), Electricity & Development in ASEAN (2005) ASEAN Energy Bull 9: 3–4
3. International Energy Agency (IEA), Energy Statistic of non-OECD countries. (2008) IEA Statistic
4. Utama NA, Ishihara KN, Tezuka T (2012) Power generation optimization in ASEAN by 2030. Energy Power Eng 4:179–314
5. Utama NA, Ishihara KN, Zhang Q, Tezuka T (2011) 2050 ASEAN electricity demand: case study in Indonesia and Cambodia, zero-carbon energy Kyoto 2010: green energy and technology. Springer, New York
6. Utama N, Ishihara K, Tezuka T, Zhang Q, Esteban M (2011) Transportation's impact assessment on construction sector. Low Carbon Econ 2(3):152–158
7. Karki SK, Mann MD, Salehfar H (2005) Energy and environment in the ASEAN: challenges and opportunities. Energy Policy 33:499–509
8. Rachmatullah C, Aye L, Fuller RJ (2007) Scenario planning for electricity generation in Indonesia. Energy Policy 35:2352–2359
9. Agya Utama N, Chayawatto N, Tezuka T, Ishihara KN, Zhang Q (2011) Power generation mixture in South East Asia region. In: Proceeding of the 4th international conference on sustainable energy and environment (SEE), Bangkok, Thailand, 27–29 February 2012

Multi-Objective Optimization Analysis of Post-Fukushima Power Generation Planning in Japan with Considering Nuclear Power's Risk Cost

Qi Zhang, Tetsuo Tezuka, and Keiichi Ishihara

Abstract In the present study, multi-objective optimization analysis was conducted on the post-Fukushima power generation planning in Japan up to 2030 from economic and environmental perspectives with considering nuclear power's risk cost. Four nuclear power scenarios were proposed in light of the nuclear accident. The obtained results show that the large scale penetration of PV (photovoltaic), wind and LNG (liquefied natural gas) power can partly replace nuclear power, however, removing nuclear power entirely by 2030 was not suggested from economic, environmental and energy security perspectives.

Keywords Economic and environmental analysis • Fukushima nuclear accident • Multi-objective optimization • Nuclear risk cost • Power generation

1 Introduction

Electricity supply in Japan is highly dependent on nuclear power, which provided 30% of their electricity demand in 2010 from a total of 54 nuclear power plants (NPP). Installed capacity of nuclear power was expected to reach 68 GW by 2030 to supply sufficient energy and reduce CO_2 emission under the strategic energy plan (SEP) released by the government in 2010 [1]. However, the Fukushima nuclear accident changed the electricity supply structure dramatically. Following on from the Fukushima Daiichi Nuclear Power Plant accident, the Japanese government is now in the throes of reviewing its nuclear power policy. At present, although it is very difficult to assess the ultimate impact on Japan's nuclear program, Japanese government has released the tentative data of economic loss of the nuclear accident and risk cost of nuclear power [2]. On the other hand, under continuing policies of

Q. Zhang (✉) • T. Tezuka • K. Ishihara
Graduate School of Energy Science, Kyoto University, Kyoto 606-8501, Japan
e-mail: ZhangQi@energy.kyoto-u.ac.jp

T. Yao (ed.), *Zero-Carbon Energy Kyoto 2012*, Green Energy and Technology,
DOI 10.1007/978-4-431-54264-3_4,

greenhouse gas reduction, it is crucial to consider scenarios for the country to realize an economic, safe and low-carbon power generation system in the future. In the present study, incorporating consideration of the attitude of the Japanese public towards nuclear power, four scenarios were proposed to incorporate effects of the accident on nuclear power policy: (1) actively anti-nuclear; (2) passively negative towards nuclear; (3) conservative towards nuclear; and (4) active expansion of nuclear power.

However, from a long-term viewpoint, apart from the safety issues of nuclear power, Japan also faces a very serious energy security problem, pressure to reduce greenhouse gas emissions and penetration of renewable energy generation bottlenecks from technological, systemic and economic perspectives. Energy supply in Japan is 96% dependent on overseas imports [3], and the price of energy resources is still increasing in international markets [4]. The domestic CO_2 emissions in Japan have increased by 20% compared to 1990 levels in the electricity generation sector up to 2009 [5] despite commitments to emissions reductions. On the other hand, the potential of renewable energy—mainly including PV and wind power—is limited in Japan due to physical-geographic reasons and constraints in technology and system integration [6]. Therefore it is crucial to reconsider the energy policy across the whole country in the mid-to-long term to realize a future clean and safe electricity system considering constraints from various perspectives.

This study presents economic and environmental analysis of the scenarios of the power generation system in Japan in light of the Fukushima Nuclear Accident up to 2030 using a multi-objective optimization method that incorporates resource availability, technology, economic and environmental perspectives. The model aims at providing decision support to select the types and total capacity of generating units to be installed and the operating levels of the various power units throughout the planning period mainly from economic and environmental perspectives considering possible nuclear power scenarios. In the long-term, nuclear power is still considered, but with advanced reactor technology to replace the current old reactors. The new technology is assumed to have legislated passive safety features that can provide coolant function using natural physical principle to avoid accidents in disasters such as the earthquake and tsunami of March 11, 2011.

2 Methodology

The long-term power generation planning model developed in the present study is a multi-objective optimization model that is mainly focusing on economic and environmental analysis of power generation system considering impacts of the Fukushima Nuclear Accident. This task can be formulated as a multi-objective problem whose solution vector is not determined by a unique combination but rather by a set of optimal arrays. The points of solution form what is commonly

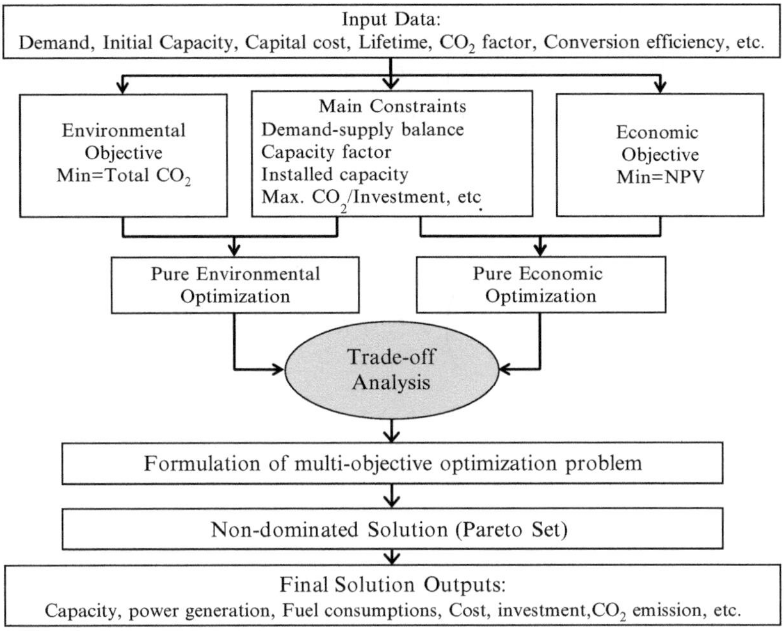

Fig. 1 Multi-objective optimization planning model for power generation system

called the non-dominated or Pareto optimal set, whose graphical representation in three dimensions is a surface, or when only two objectives are involved, it generates a trade-off curve that constitutes the range of optimal choices available for planners. For each one of the Pareto arrangements, it is impossible to improve one objective without worsening another. The determination of the optimal arrays for planning an integrated energy system is basically a non-linear programming problem that can be approached with several numerical techniques. The general flow chart of the multi-objective optimization model is illustrated in Fig. 1. By integrating the necessary input information, the model can provide decision support to planners by rationalizing the comparison among different alternative solutions, thus enabling the planners to grasp the inherent conflicts and trade-offs among the distinct objectives for selecting a satisfactory compromise solution from the set of non-dominated solutions.

The economic and environmental objective functions are shown in Eqs. (1)–(6). There are still many constraints from supply–demand balance, installed capacity change, investment, CO_2 emission, etc. that are not listed here. The model was developed by using GAMS (General Algebraic Modelling System) [7].

$$\min \ NPV = \min(TCC + TFC + TOMC - TRC) \tag{1}$$

NPV = the net present value of total cost for power generation;
TCC = the total capital cost of the new power plants in the planning period;

TOMC = the total operation and management cost;
TFC = the total variable cost including fuel cost and management cost;
TRC = residual values of the new power plants at the end of the planning period.

The detailed calculations of the four kinds of cost are provided as below.

$$TCC = \sum_{y=2}^{Y} \left\{ \left[\sum_{i=1}^{m} (newpp_{y,i} \times cc_{y,i}) \right] \times (1+R)^{-5(y-1.5)} \right\} \tag{2}$$

$$TFC = \sum_{y=1}^{Y-1} \left\{ \left[\left(\sum_{i=1}^{m} (fs_{y,i} \times fc_{y,i}) \right) \times (1+R)^{-5(y-1)} + \left(\sum_{i=1}^{m} (fs_{y+1,i} \times fc_{y+1,i}) \right) \times (1+R)^{-5(y)} \right] \div 2 \times 5 \right\} \tag{3}$$

$$TOMC = \sum_{y=1}^{Y-1} \left\{ \left[\left(\sum_{i=1}^{m} (pp_{y,i} \times omf_{y,i} + rc_{y,i}) \right) \times (1+R)^{-5(y-1)} + \left(\sum_{i=1}^{m} (pp_{y+1,i} \times omf_{y+1,i} + rc_{y+1,i}) \right) \times (1+R)^{-5(y)} \right] \div 2 \times 5 \right\} \tag{4}$$

$$TRC = \sum_{y=2}^{Y} \left\{ \left[\sum_{i=1}^{m} \left(newpp_{y,i} \times cc_{y,i} \times \frac{(Y+0.5-y) \times 5}{lf(i)} \right) \right] \times (1+R)^{-5(y-1.5)} \right\} \tag{5}$$

Where

Y = the number of the year spans in the planning period;
y = each year span;
i = a specific type of power plant;
m = number of the types of power plants;
$newpp_{i,y}$ = install capacity of new power plant "i" in the year "y";
$cc_{i,y}$ = capital cost per unit installed capacity of type "i" in the year "y";
R = discount rate;
$pp_{y,i}$ = install capacity of power plant "i" in the year "y";
$omf_{y,i}$ = operation and management cost coefficient of power plant "i" in the year "y";
$rc_{y,i}$ = risk cost of power generation technology i in the year y;
$fs_{i,y}$ = fuel supply of power plant type "i" in the year "y";
$fc_{i,y}$ = fuel cost per unit fuel supply of power plant type "i" in the year "y";
lf_i = lifetime of power plant type "i";

In the present model, every year span includes 5 years.

The environmental optimization objective function is shown in below:

$$\min Total\ CO_2 = \min \sum_{y=1}^{Y-1} \left\{ \sum_{i=1}^{m} \left[(fs_{y,i} + fs_{y+1,i}) \times CO_2 f_i \div 2 \times 5 \right] \right\} \quad (6)$$

Where, $CO_2f_i = CO_2$ emission factor per unit fuel type "i"

3 Data

There are many arguments about the future development of nuclear power in Japan as a result of the various conflicting objectives. Therefore, four nuclear power development policy scenarios are proposed: (1) actively anti-nuclear; (2) passively negative towards nuclear; (3) conservative towards nuclear; and (4) active expansion of nuclear power. In the actively anti-nuclear scenario (N1), all the NPPs will be closed permanently by 2015, and all NPPs under construction and in planning will be cancelled; in the passively negative nuclear scenario (N2), the existing NPPs in operation will be closed at the end of their lifetime, 40 years; in the conservative nuclear scenario (N3), comparing with "N2", all NPPs under construction and in planning will be continued but with 5 year postponement from the currently proposed schedule; finally in the active expansion nuclear power scenario "N4", nuclear power is developed according to the government plan published in 2010 with the exception of Fukushima Daiichi 1–4 plants being closed in 2011 permanently. According to the nuclear power's historical installed capacity increases and future development plan [1], the installed capacities to 2030 in the four scenarios are shown in Fig. 2.

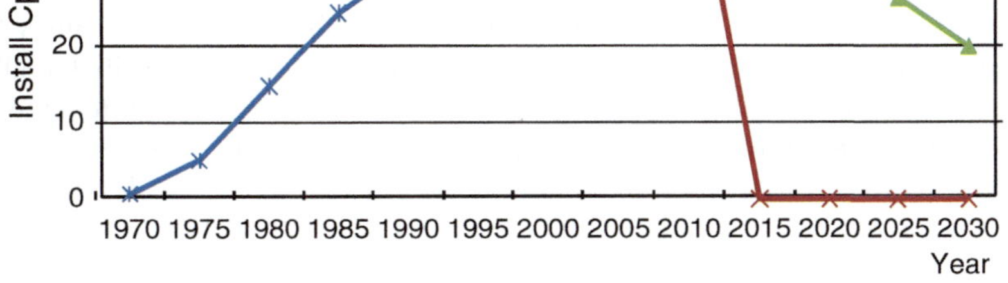

Fig. 2 Four proposed nuclear power policy scenarios in Japan

Table 1 Economic data for optimization analysis of power generation system planning

	Capital cost (1,000 yen/kW)	Fuel cost (yen/MJ)	Annual OM cost/capital cost (%)
Nuclear	350		5
Coal	230	0.35	5
LNG	160	0.57	5
Oil	190	0.91	5
Biomass	500	0.57	5
PV	500	0	1.5
Wind	200	0	2
Hydro	850	0	1
Pumped hydro	850	0	1

The basic cost information for various power generation technologies is shown in Table 1 [2]. For nuclear power, fuel recycle costs of about 1.5 yen/kWh also need to be considered, and the capital cost includes not only building cost but also decommissioning cost. However, the future price of energy resources in the global market is quite uncertain (although likely to rise), and the given cost information is used as an example. Following on from the March 2011 Fukushima Daiichi accident, new cost estimates for nuclear power generation in Japan have been conducted. The estimate factors in the possibility of a core meltdown at a nuclear plant, and includes societal costs such as the cost of dealing with future accidents and policy costs related to governmental budgets for the first time. Particularly, the newest estimate includes 0.5 yen/kWh for dealing with future nuclear risks as shown in Eq. (7). The 0.5 yen/kWh for future nuclear risk is a minimum: the cost would increase by 0.1 yen/kWh for each additional 1 trillion yen of damage [2]. Furthermore, the discount rate is stipulated as 3% in the present study.

$$rc_y = \frac{total\ Economicloss}{\sum_y nuclear\ pg_y} \tag{7}$$

The CO_2 emissions factors of various power generation technologies are shown in Fig. 3 [8], in which emissions in manufacture processes are not included. According to the calculations, in 2030, most coal-fired power plants will be ultra-supercritical technology with >600°C steam temperature, and LNG-fired power plants will be combined-cycle plants (rather than open cycle) in Japan. CO_2 emissions could be reduced by about 5% and 20%, respectively, due to the efficiency improvements in advanced coal-fired and LNG-fired power generation technologies.

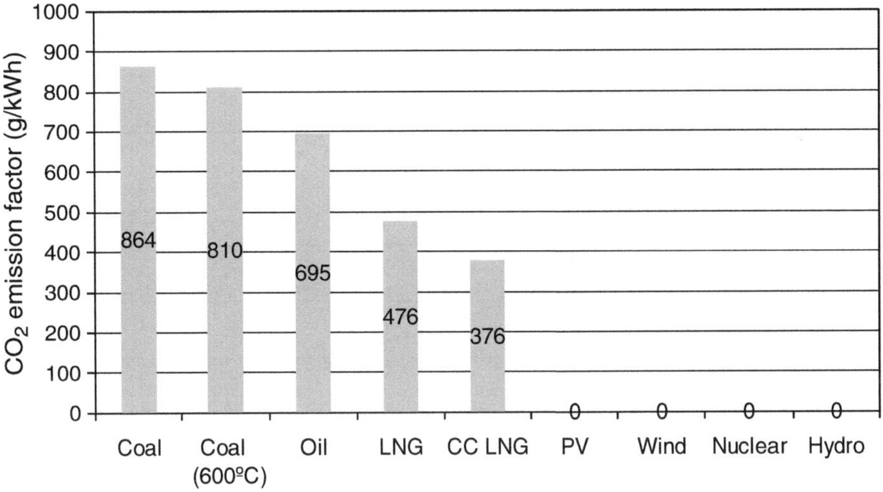

Fig. 3 CO_2 emissions factors of various power generation technologies

4 Results

The multi-objective optimization solutions of the electricity generation planning based on the proposed four nuclear power scenarios is shown in Fig. 4, generated by possible non-inferior solution points (or optimal operational strategies) for the competing objectives including NPV and CO_2 emissions, in which the two endpoints of individual curve are pure least cost and least CO_2 emission optimization results respectively. In the trade-off curve of scenario 1 as an example, the point “G” is the pure least cost solution and point “A” is the least CO_2 emission solution. The comparison of the curves of the four scenarios shows that with the increase of nuclear power in the generation systems, both absolute values and variable ranges of NPV and CO_2 emissions become smaller. In scenario 1 (actively anti-nuclear), the NPVs change from 115 to 135 trillion yen with 20 trillion yen difference and the CO_2 emissions change from 8,500 to 10,000 million tonnes with 1,500 million tonnes difference in least cost and least CO_2 emission scenarios. However, in the scenario 4 (active expansion of nuclear power), the NPVs change from 95 to 105 trillion yen with 10 trillion yen difference and the CO_2 emissions change from 6,000 to 5,100 million tonnes with 900 million tonnes difference in least cost and least CO_2 emission scenarios.

In order to understand that each optimization point has a set of different power generation expansion plan, the electricity mixes in 2030 of the scenario 1(S1) from “A” to “B” are illustrated in Fig. 5. Each trade-off point corresponds to an optimal solution of the generation expansion. For example, as the degree of satisfaction of the economic objective increases (from “A” to “B”), the total CO_2 emissions in 20 years increases from 8,500 million tonnes to 8,700 million tonnes, as the share of

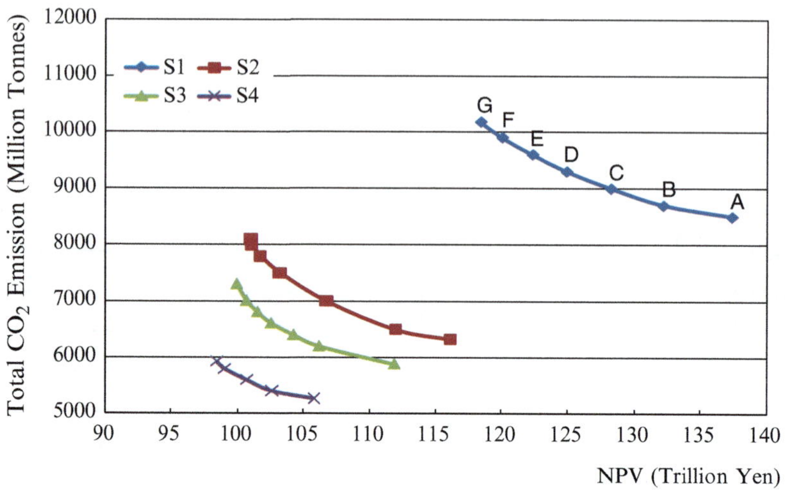

Fig. 4 Economic and environmental multi-objective trade-off analysis results of the four scenarios

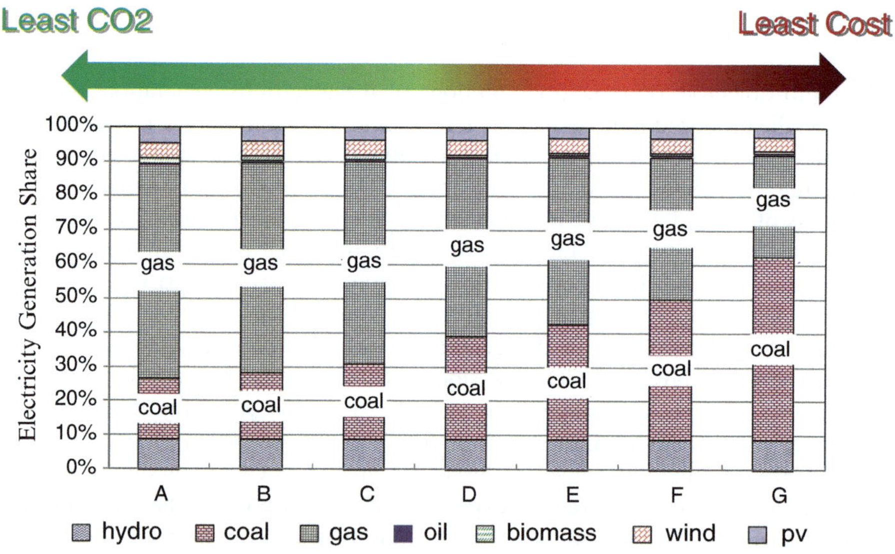

Fig. 5 Electricity mixes in 2030 for the seven optimization points of the scenario 1

coal-fired power increases from 15% to 20%. Comparing a shift from A to G, the total CO_2 emissions in 20 years increases from 8,500 million tonnes to 10,200 million tonnes, as the share of coal-fired power increases from 15% to 65%, and gas-fired power decreases from 65% to 30%. The trade-off curve of

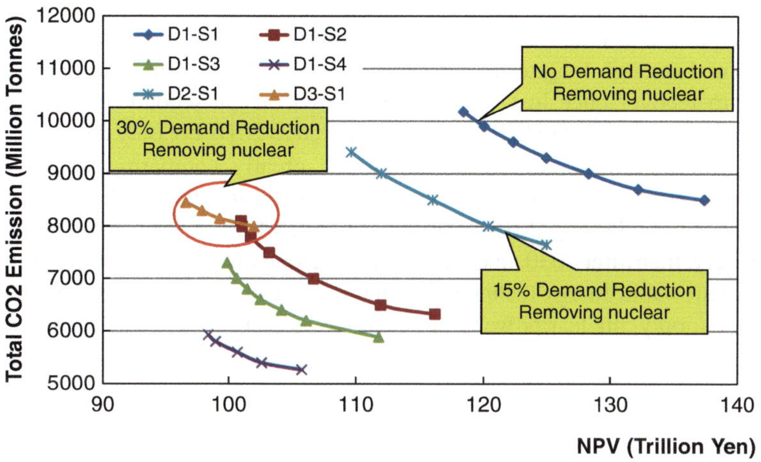

Fig. 6 Multi-objective optimization trade-off results with demand reduction

economic and environmental multi-objective optimization can be used to determine reasonable CO_2 emissions taxes or clean energy subsidies, and can also be used to conduct analysis on economic environmental caps for the power generation system in the future study. On the other hand, Japan highly depends on overseas imports of energy resource. It is very obvious, in the first nuclear power policy scenario (S1), much more LNG or coal need to be imported from overseas, therefore, the energy security problem will become very serious.

If the demand can be reduced by various electricity saving methods, optimization result is expected to be changed. Three demand scenarios were proposed: no electricity saving in scenario D1; 15% demand reduction in scenario D2; and 30% demand reduction in D3. As shown in Fig. 6, if the electricity demand can be reduced by 30%, removing nuclear power entirely by 2015 is economic and environmental acceptable comparing with other nuclear scenarios. However, even 15% reduction is very difficult if there is no significant changes in people lifestyle and industry structure in Japan.

Apart from economic and environmental aspects, nuclear power safety issue should be focused. Nuclear power safety does not only include operation of nuclear power plant but also the storage, transportation, recycle and disposal of spent nuclear fuel. At present, there is about 20,000 tonnes spent nuclear fuel stored inside of nuclear power plants or interim storage facilities in Japan. The management and disposal of the spent nuclear fuel requires nuclear power experts, various advanced technologies and large financial support. If nuclear power is removed entirely by 2015 in Japan, there will be a very serious problem of the lack of skilled staff to operate nuclear power plants and manage spent nuclear fuel in the near future, and it will be very difficult to save and develop many advanced and important technologies of nuclear power.

Therefore, removing nuclear power entirely by 2015 can't improve the safety of nuclear power. It will lead to the losses of important experts and technologies which are obtained through long term and large financial investments, and it will be very difficult to recover once lost. On the other hand, in the second scenario in which nuclear power is reduced to half in 2030 compared to 2010, not only the economic and environmental performance can be improved greatly comparing with the 100% removing scenario, it is also a practical way to ensure the nuclear power safety comprehensively. During 2010–2030, the storage of spent nuclear fuel can be disposed safely, then after 2030, nuclear power development policy needs to be evaluated again. If renewable energy technology can provided sufficient, reliable and economic electricity supply, nuclear power can be reduced to zero by 2050 at that time.

5 Conclusion

In the present study, multi-objective optimization analysis was conducted on power generation planning from economic and environmental perspectives in Japan considering nuclear risk cost in light of Fukushima Accident. Four nuclear power scenarios were proposed: (1) actively anti-nuclear; (2) passively negative towards nuclear; (3) conservative towards nuclear; and (4) active expansion of nuclear power. The obtained results showed that removing nuclear power entirely may lead to the increases of 40 trillion yen cost, 5,000 million tonnes CO_2 emissions totally. Furthermore, the economic and environmental multi-objective trade-off curves showed that penetration of nuclear power can reduce the values and variable ranges of economic and environmental performances effectively. Although, 30% demand reduction can make removing nuclear power economic and environmental acceptable, big demand reduction can't be expected very much if there is no significant changes in people lifestyle and industry structure. Based on the obtained results, the study shows that the large scale penetration of PV, wind and LNG power can replace a part of nuclear power in future power system in Japan by 2030, but it is very difficult to remove nuclear power entirely before 2030 considering the impacts from economic, environmental, energy security and nuclear safety perspectives.

Acknowledgment The authors would like to thank the GCOE program in Graduate School of Energy Science, Kyoto University and Kan Gen Kon for the great support and helpful comments.

References

1. METI (2010) The strategic energy plan of Japan, 2010. Available at http://www.meti.go.jp/english/press/data/pdf/20100618_08a.pdf, Accessed May 2012
2. Cost Calculation Council, National Policy Unit, Japan (2011) Cost estimation of various power generation technologies. Available at http://www.npu.go.jp/. Accessed Dec 2011 [in Japanese]
3. IEA (2010) Energy balance of OECD countries. Paris, 2012

4. IEA (2010) Energy price and taxes, quarterly statistics, first quarter 2011. Paris, 2011
5. Greenhouse Gas Inventory Office of Japan (1990–2009) The GHG emissions data of Japan, April 2011. Available at http://www-gio.nies.go.jp/index.html, Accessed May 2012
6. ME (Ministry of Environment of Japan) (2011) Study of potential for the introduction of renewable energy (FY 2010) [in Japanese]
7. GAMS (2008) A user's guide. Available at http://www.gams.com, Accessed May 2012
8. Imamura E, Nagano K (2010) Evaluation of life cycle CO_2 emissions of power generation technologies, CRIEPI Report, p 34 [in Japanese]

(iii)
Session Papers

Thailand's Security of Energy Supply: Import Dependency Vulnerability Assessment

Kamonphorn Kanchana and Hironobu Unesaki

Abstract This paper assesses Thailand's security of energy supply by analysing how it has been evolved over past two decades and by applying a composite indicator in order to perform a comprehensive supply vulnerability assessment. The results show low diversity in the country's energy mix, which is, by half, crude oil concentrated. Market concentration is mainly based on suppliers from the Middle East, which leads to transportation risks due to the number of chokepoint shares. On contrary, coal and natural gas imports is from countries in South East Asia. Regarding indigenous as well as intraregional resource potential, natural gas is considered as better energy source option in terms of import dependency vulnerability. Encouragement of natural gas usage in transport sector will lead to lesser dependency on imported oil supply.

Keywords Energy security indicator • Import dependency • Supply vulnerability

1 Introduction

According to economic growth, Thailand's energy demand and import have increased significantly. From 1971 to 2009, energy demand increased by the annual average rate of 5.6% while net energy import grew at the average rate of 5.7%. During the same period, the average ratio of net energy import to energy supply in the economy was 46.3% [1]. Due to gradual growth in energy demands, increasing reliance on import energy supply and limited resource availability, energy security is a major challenge. Thai energy policy is much driven by external factors, particularly tight oil market condition. The country's energy security policy is to enhance energy development for greater self-reliance with a sufficient and stable

K. Kanchana (✉) • H. Unesaki
Graduate School of Energy Science, Kyoto University, Kyoto 606-8501, Japan
e-mail: kamonphorn.kanchana.47n@st.kyoto-u.ac.jp

T. Yao (ed.), *Zero-Carbon Energy Kyoto 2012*, Green Energy and Technology, DOI 10.1007/978-4-431-54264-3_5,

energy supply by (1) maximizing indigenous resources, (2) exploring potential energy resources domestically and internationally, (3) developing appropriate energy mix based on a variety of sustainable supply, (4) promoting the use of alternative energy and renewable energy sources, and (5) diversifying energy sources and types to reduce dependence vulnerability [2].

Still, without master plan, energy security policy direction is not clear. Moreover, there are only a small number of existing energy security assessment studies, especially on import risk assessment. Most available indices only provide statistic backgrounds. As a result, the objective of this paper is to assess Thailand's security of energy supply emphasizing on the linkage of energy import dependency, the degree of diversification, geographical risk, etc. in order to identify vulnerabilities or key challenges that may affect energy security. Following this introduction, the second part of the paper introduces the methodology while the next section presents assessment results and explanation and the paper concludes with remarks on future energy security policy planning.

2 Methodology

This paper assesses Thailand's energy security in terms of import dependency vulnerability based on primary energy data and statistics gathered from the International Energy Agency, the Ministry of Energy, the Ministry of Commerce and the Customs Department within the research time frame of 1990–2010.

The assessment index used in this paper is resource concentrated in which primary energy supply (PES) is divided into four categories including coal (coal and coal products), oil (crude oil and oil products), natural gas and renewable energy sources (biofuel, biomass, hydropower, etc.). A composite indicator was built based on the adaptation of three aggregated indices proposed by Gupta's Oil Vulnerability Index (OVI) [3], European Commission JRC's indicator for energy security [4], and Murakami et al.'s energy security assessment model [5].

The assessment can be divided into three main parts. The first part, the security of supply, focuses on the security of supply in general by assessing (1) the degree of diversification of PES using the Shannon–Wiener Index (SWI), (2) the country's domestic resource using self-sufficiency level, and (3) overseas resource reliance using import dependency rate. The second part focuses on energy import vulnerability assessment based on geographical risks by measuring (1) energy market concentration using Herfindahl–Hirschman index (HHI) and (2) transportation risks using chokepoint and critical point shares. The computation in this part includes 50% of imports from Indonesia to be accounted for Malacca chokepoint share. A composite indicator showing import vulnerability comparison between coal, crude oil and natural gas consists of six indicators including source share in total primary energy supply, self-sufficiency, import dependency diversification of import source by region, dependency on country ranked more than five based on OECD country risk [6], and chokepoint diversification.

The value of SWI used to measure diversity scale ranges from near 0 (low richness and evenness). For market concentration, the HHI value lower than 0.15 indicates unconcentrated market while the value of 0.25 or greater indicates high concentration. OECD country risk ranges from 0 (no risk) to 7 (highest risk).

3 Assessment Results

3.1 *The Security of Supply*

3.1.1 Energy Resource Diversity

As shown in Fig. 1, the degree of Thailand's primary energy supply diversity based on Shannon–Wiener Index (SWI) for diversity has the average of 1.24, which refers to low richness and evenness, due to the fact that the country's energy mix is, by half, crude oil-concentrated.

3.1.2 Domestic Energy Resources

Energy self-sufficiency reflects the presence of domestic energy resources in supply structure. Low primary energy self-sufficiency rate implies the reliance on overseas sources. Domestic energy production accounts for averagely 58.51% of total primary energy supply where renewable energy is completely indigenous. Although Thailand has certain domestic crude oil resources, more than 70% of crude oil consumption is imported. Its self-sufficiency rate is thus remarkably low (the average rate of 21.7%) comparing to continuingly increasing crude oil demands (average annual growth rate of 7.1%).

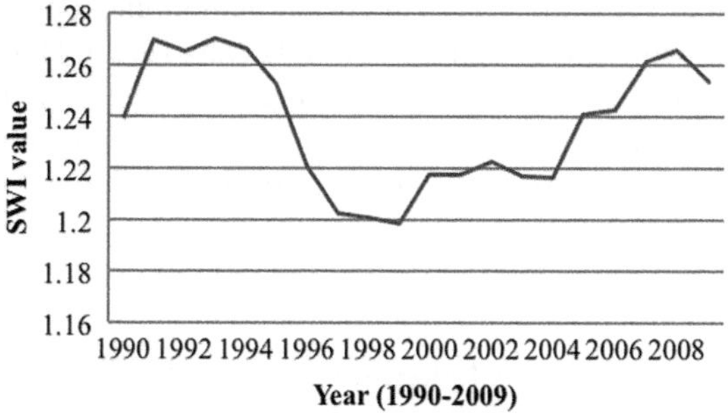

Fig. 1 SWI for the degree of energy resource diversity

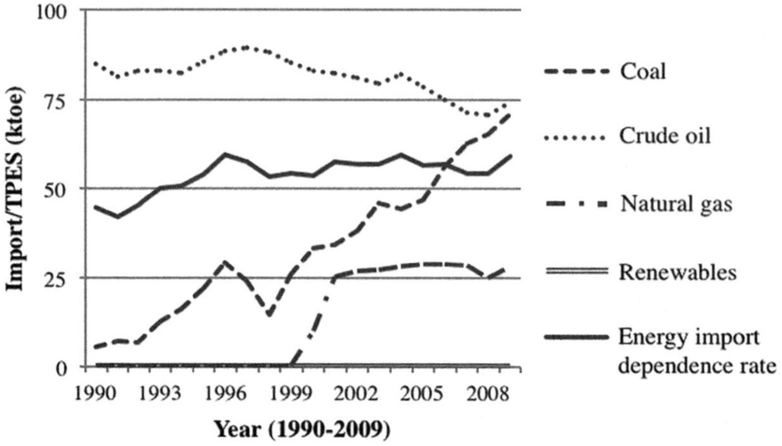

Fig. 2 Energy import dependence rate

3.1.3 Overseas Energy Resources

In the past 15 years, more than 50% of Thailand's domestic energy use depended on overseas sources, particularly crude oil. Unlike renewable energy, quantities of imports of coal and natural gas have risen since 1999, as there have been increasing demands in power generation, industry and transport sectors. Yet, crude oil import dependency rate remains the highest of all resources with the average of 81.43% as shown in Fig. 2.

3.2 Energy Import Vulnerability Assessment

3.2.1 Energy Market Concentration

The following values show the degree of energy geographical market concentration based on the Herfindahl–Hirschman index (HHI): 0.54 for crude oil, 0.55 for natural gas and 0.70 for coal. The results indicate high market concentration. Figure 3 below shows energy import market shares in which 71.89% of crude oil is imported from the Middle East whereas more than two-third of coal and natural gas imports are from countries in South East Asia.

3.2.2 Transportation Risks: Chokepoints and Critical Points

Related to previous indicator, the implication of market concentration is the geographical risk assessment due to transportation risks. Thailand's chokepoint and critical

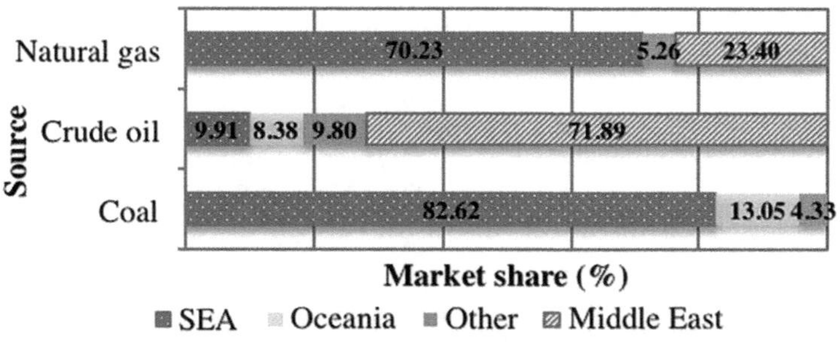

Fig. 3 Energy market share by source and region

Table 1 Chokepoint and critical point assessment

Topic	Crude oil	Coal	Natural gas
SWI for chokepoint/critical point diversification	0.88	0.03	1.06

point diversification is very low. As shown in Table 1, the SWI chokepoint diversity level for crude oil is 0.88 as a result of high dependence on Malacca Strait (75.52%) and Hormuz Strait (57.27%). Similarly, coal chokepoint diversification is very low as two-third of net coal import is from Indonesia. Malacca Strait is thus the most significant chokepoint for coal and crude oil imports. However, transportation vulnerability for natural gas is not maritime chokepoint but critical infrastructure like Thailand–Myanmar and Thailand–Malaysia transnational gas pipelines. Almost all of the imports with no chokepoint share are intra-regional.

3.2.3 Energy Import Vulnerability Assessment by Sources

Presented in Fig. 4, crude oil is the most vulnerable energy source due to its high share in the total primary energy supply, self-sufficiency rate, import dependency and its geographical risks from counterpart diversity. It is thus a major challenge in energy security balance.

Unlike crude oil, Thailand has significant natural gas production capability so its share of net import is only 11.9% while the self-sufficiency rate is 71.9%. Since more than two-third of natural gas consumption is from domestic production, import dependency and import source diversification is very low (0.83). Yet, the vulnerability is that natural gas major supplier is from Myanmar, which is ranked as level 7 based on OECD country risk classification (the most risky for investment), which threatens import security.

Coal, however, shows high import dependency. Due to its small share in TPES and domestic amount of coal reserves, the degree of diversification of coal suppliers is low. Most imports are from South East Asia countries. As a result, coal import vulnerability is not an outstanding threat to energy security.

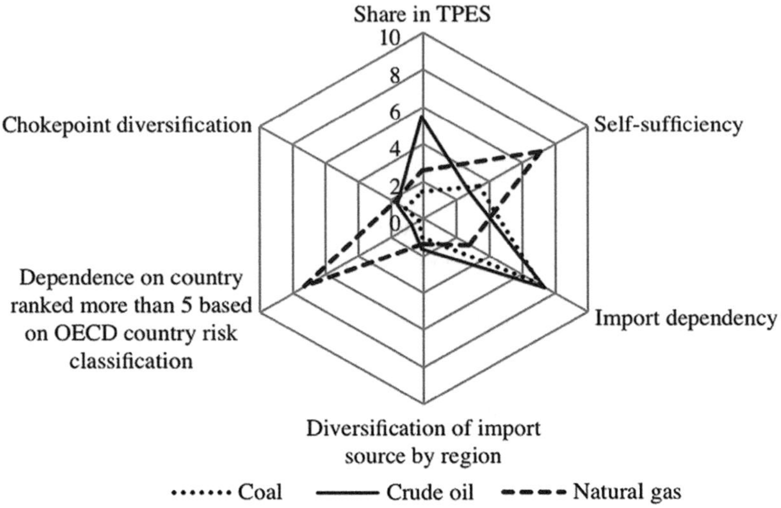

Fig. 4 Energy import vulnerability assessment by source

Comparing to crude oil and coal, natural gas would be a better energy source. Encouragement of natural gas usage in the transport sector and the use of electric vehicles will lead to lesser dependency on imported oil supply. In addition, natural gas availability in the region will be able to leverage limited natural gas reserves.

4 Future Policy Planning

Considered from energy import vulnerability assessment results, Thailand's current energy security strategy is arbitrary. Despite the attempts to encourage the production and consumption of domestic energy sources to ensure energy security, energy import dependency level is considerably high. Although the government aims to diversify energy sources and suppliers to decrease import reliance, geographical risk assessment shows heavy reliance from crude oil import from the Middle East, which leads to transportation vulnerability. Diversification of import source by regions and transportation routes should be taken into account.

To ensure secure energy availability, Thailand should have governmental emergency oil stockpiling instead of a stockholding from private sector, which will be sufficient only for 60 days of domestic usage. Even though global oil market ensures the availability of suppliers, governmental strategic oil stockpiling for emergency is indispensible. Apart from supply security, the control of energy demands should also be prioritized.

As mentioned that natural gas can be a good energy option, regional cooperation on natural gas exploration, investment, and trade should be encouraged in both public and private sectors. In addition, due to the vulnerability of intraregional natural gas transportation, the government should also consider transnational energy infrastructure security. The next research studies further into pipeline security and regional cooperation in ASEAN.

Acknowledgment The authors were grateful for the support of the Ministry of Education, Culture, Sports, Science and Technology of Japan via "Energy Science in the Age of Global Warming" of Global Center of Excellence (G-COE) program (J-051).

References

1. Ministry of Energy. (2011). Thailand energy situation report. Retrieved from http://www.dede.go.th
2. Thailand's Parliament. (2011). Policy announcement. Retrieved from http://www.cabinet.thaigov.go.th
3. Gupta, E. (2008). Oil vulnerability index of oil-importing countries. Energy policy, 36, 1195–1211
4. Join Researc Center (JRC), European commission. (2011). Energy security indicators. Retrieved from www.jrc.ec.europa.eu
5. Murakami, T., Motokura, M., Kutani, I. (2011) An analysis of major countries' energy security 165 policies and conditions. IEEJ (March 2011), 1–36.
6. OECD (2011) Country risk classifications of the participants to the arrangement on officially supported export credits. Retrieved from http://www.oecd.org/tad/exportcredits

Measures to Promote Energy Conservation in Indonesian Households with Different Cultural Backgrounds: An Analysis on Electricity Prices Perspective

Muhammad Ery Wijaya and Tetsuo Tezuka

Abstract The variety of cultural backgrounds play a huge role on the residents' perspective on electricity price, and that will eventually lead to the decision-making behaviour regarding the electricity consumption. This study, under the framework analysis of different cultural backgrounds and ethnicities in Indonesia are represented by two areas—Yogyakarta as the representative of Javanese culture and Bandung as the representative of Sundanese culture, seeks to the following: (1) to analyze and compare the understanding of electricity price perspective and its changes in those societies, and (2) to analyze and compare the effect of monthly payment systems in awareness of electricity consumption between two societies. The result of study is expected to design a decently compiled strategy to improve energy conservation effort in Indonesia related to the electricity price perspective in local societies. The questionnaire survey method is employed to obtain information on electricity prices perspective. The results show that people in Bandung had significantly better understanding of electricity price as compared to people in Yogyakarta. However, change of electricity price does not serve as a signal in decision-making about electricity consumption in the homes of both cities. In order to improve the awareness of energy consumption, the pre-paid system of electricity payment should be widely introduced in Yogyakarta. On the flip side, the electricity payment with post-paid system is more suitable for Bandung.

Keywords Culture • Electricity price • Energy conservation • Household • Indonesia

---

M.E. Wijaya (✉) • T. Tezuka
Energy Economics Laboratory, Department of Socio-Environmental Energy Science, Graduate School of Energy Science, Kyoto University, Kyoto 611-0011, Japan
e-mail: muhammad.wijaya.23v@st.kyoto-u.ac.jp

T. Yao (ed.), *Zero-Carbon Energy Kyoto 2012*, Green Energy and Technology, DOI 10.1007/978-4-431-54264-3_6,

1 Introduction

Java, Madura and Bali Islands are identified as the biggest electricity consumer in Indonesia with the largest share of electricity consumer is the industrial sector at nearly 47%, followed by the household sector with 39% [1]. According to National Electricity Master Plan 2008–2027, the household sector is predicted to become the biggest electricity consumer in the future. Households are thus an important group when it comes to addressing energy conservation in Indonesia.

Indonesia is a country where the society consists of different cultural backgrounds or ethnicities that are distinguished into several regions. Java Island as the centre of economic and development, is home to two largest ethnic groups in Indonesia namely the Javanese and the Sundanese, accounted for 41.7% and 15.4% of national population, respectively [2]. In addition, Javanese and Sundanese people also counted as the largest electricity consumers in this area. Therefore, the variety of cultural backgrounds in Indonesian society may affect the residents' behaviour in electricity consumption. A thorough understanding on the characteristic in electricity consumption is important to design an energy conservation policy based on local cultures to improve the success of policy implementation.

One of the variables that is often considered as an important indicator in decision-making when people consuming electricity is electricity price [3]. Therefore, this study aims: (1) to analyze and compare the understanding of electricity price perspective and its changes in Javanese and Sundanese societies; and (2) to analyze and compare the effect of monthly payment systems in awareness of electricity consumption between two societies. The results of study is expected to compile a strategy to improve energy conservation effort in Indonesia related to the electricity price perspective in local societies.

2 Objects of Study

An ethnic group refers to a cultural identity that takes into account language, tradition and patterns of behaviour. Javanese is typically known lives in modest and practical lifestyle, while Sundanese lives more materialistic and consumerist as open minded society [2, 4]. As the largest population groups in Indonesia, Javanese and Sundanese societies could be considered as the biggest objects to promote energy conservation in the residential sector. These differences in the way of thinking in between both of the societies could affect the perspective on electricity consumption, especially on the people perspective on electricity price. Therefore, two areas are selected as objects for this study: Yogyakarta as representative of Javanese culture and Bandung as representative of Sundanese culture. These cities were selected not only due to the largest electricity consumption in the country, but also due to cultural density that each city represents for its respective culture.

3 Survey Method

This study obtained primary information by using questionnaire survey conducted in two cities. Due to time constrains, the number of respondents is limited to 100 households randomly selected to be investigated in-depth from each city. The research was conducted from October to November 2011. There are about 30% of households from both cities in this survey using prepaid system in their electricity payment.

The questionnaire is designed to identify the people's perspective on electricity price and electricity monthly payment mechanism in Indonesia to the awareness of electricity consumption. The questionnaire follows the scoring system based on Likert scale to ease the respondents in rating from many options of the answer level of agreement or disagreement. The questions were consisted of the following items: (1) knowledge in unit of electricity prices; (2) awareness of the changes of monthly electricity bill; (3) response in the change of electricity price; and (4) effects of electricity payment system. After receiving the responses from respondents, the average scores given by respondents for every question are then calculated. In order to get a comparison analysis of decision-making in the purchase and use of electrical appliances in two cities, a statistical analysis called T-test is employed.

4 Results and Discussion

4.1 Socio-Economic Profile of Respondents

Majority of the respondents in both cities are male, and most of the respondents hold a senior high school diploma. The average age was 45 years in Yogyakarta and 43 years in Bandung. The average family size for each respondent is four in both cities. In Yogyakarta, approximately 75% of those surveyed have incomes lower than US$ 300. In Bandung, the income distribution is also dominated by those with incomes less than US$ 300, but the proportion, more than 46% of the total respondents, is considerably lower than that were found in Yogyakarta. Generally, both cities show similar trends in income level because the income distribution is dominated by the households with the lowest income level. However, the households in Bandung have higher income average than the households in Yogyakarta.

4.2 Knowledge and Awareness in Electricity Price

In this survey, the respondents' knowledge and perspective on electricity price as set by government is investigated. The first question is by evaluating on how familiar they are with the unit price per kWh of their electricity bill. These ratings

Table 1 The comparison between Yogyakarta and Bandung of respondents' responses about their understanding of the price of electricity

Question	City	Mean	*p*-value*
How do you understand with the electricity prices?	Yogyakarta	2.15	0.02
	Bandung	2.42	
How do you aware of changes in monthly electricity bill?	Yogyakarta	3.69	0.09
	Bandung	3.50	

*p-value <0.05 is significantly different

use a scale of 1 (not at all) to 4 (highly). This same topic was also examined by asking the respondents about their awareness of the changes in their monthly electric bill. These ratings use a scale of 1 (not at all) to 5 (highly). This assessment is required to furnish an explanation of the responses from the respondents in Yogyakarta and Bandung about the energy consumption of appliances.

The results (see in Table 1) indicate that people in Bandung have a significantly better understanding of the unit price of electricity than do the people in Yogyakarta, since people characteristic in Bandung is more attentive and knowledgeable towards their consumption, while for Javanese people, understanding unit electricity price seems complicated. In contrast, no significant difference is found between Bandung and Yogyakarta respondents in their awareness of changes in their monthly electric bill. The average frequencies of the respondents' responses about electrical prices differ amongst the cities. However, the results on the four-point scale show that the price of electricity is not well understood by the respondents in Bandung and Yogyakarta. In contrast, the awareness of the respondents in both cities of the changes in their monthly electric bill is fair to very high. It reflects that although the cultural background of both societies is quite different, the total monthly electricity consumption bill brings people to higher awareness level in their consumption despite their knowledge of the prices of electricity set by the government is sparse.

4.3 Effects of Change in Electricity Prices

The change in prices would be assessed to see the response of consumers in decision-making of eight frequently used devices: washing machines, water pumps, rice cookers, lighting devices, electric fans, air conditioners, refrigerators, and television sets. Respondents in each city are asked on how their usage of each appliance would change if the electricity price increased by 10% from current price. The question is scaled from one (large decrease) to five (large increase).

The results indicate that respondents in Yogyakarta tend to reduce their electrical appliances usage with significantly differences than in Bandung for TV, refrigerator, rice cooker, water pump, and washing machine. However, there are no significant differences found between Bandung and Yogyakarta for air conditioner,

Table 2 The comparison between Yogyakarta and Bandung of respondents' responses in change of electricity prices

Devices	City	Mean	*p*-value*
TV	Bandung	2.77	0.04
	Yogyakarta	2.62	
Refrigerator	Bandung	2.99	0.00
	Yogyakarta	2.76	
Air conditioner	Bandung	2.19	0.76
	Yogyakarta	2.10	
Electric fan	Bandung	2.85	0.06
	Yogyakarta	2.69	
Lighting	Bandung	2.70	0.67
	Yogyakarta	2.67	
Rice cooker	Bandung	2.96	0.00
	Yogyakarta	2.70	
Water pump	Bandung	2.86	0.01
	Yogyakarta	2.63	
Washing machine	Bandung	2.98	0.00
	Yogyakarta	2.61	

*p-value <0.05 is significantly different

electric fan and lighting. Details of the comparison of respondents' response in change of electricity prices in both cities can be seen in Table 2.

Although there are several significant differences, in the point of view of the average frequency of respondents' response, if the electricity price is increased by 10% the people in Bandung tend not to change their usage of appliances and keep it normal (all average frequency of respondents' response are over than 2.5), except for air conditioners use that expected to be slightly decreased. Similar situation to Bandung, respondents in Yogyakarta also tend to keep electricity use normal, except for their usage of air conditioner. From these findings, change of electricity prices does not serve as a signal in decision-making about electricity consumption in the home.

The results of this question has clarified that policy option to increase electricity price in order to increase people's awareness in their electricity consumption will not result in a significant reduction in electricity consumption in both cities, although people in Bandung has better understanding of electricity prices. This rationale is due to the fact that electricity is one of primary necessities of life and is routinely used by people in their daily life. In the case of basic necessity, the cultural differences either modest or consumerist lifestyle could not affect to their consumption pattern.

4.4 Effect of Monthly Payment Systems in Awareness of Electricity Consumption

In Indonesia, there are two payment systems that have been enacted, namely post-paid system and prepaid system. The post-paid system charges consumer with both load and usage fees. The consumers who adopt this system use electricity in advance and pay their consumption by the end of each month. On the other hand, by using

Table 3 The comparison between Yogyakarta and Bandung on effects of electricity payment systems to monthly electricity use

City	Payment system	Mean	p-value*
Yogyakarta	Post-paid	4.14	0.63
	Prepaid	4.23	
Bandung	Post-paid	4.37	0.01
	Prepaid	3.97	

*p-value <0.05 is significantly different

prepaid system, the consumers are free of charge from a load fee. But to be able to use electricity, consumers should purchase a voucher in advance through specific designated kiosks with an amount of credits desired by the consumers themselves. The prepaid system is able to show the amount electricity consumed in real time, and the remaining of electricity that can be used by consumer. Consumer must buy additional voucher should their credit runs out. Thus, the next assessment is to see the awareness of respondents in their monthly electricity use between prepaid and post-paid systems on a scale of 1 (not at all) to 5 (highly). The results are then presented in Table 3.

In Yogyakarta, there is no significant difference of respondents' awareness in their electricity payment between the two systems. However, the respondents with prepaid system in Yogyakarta show better awareness in electricity payment than respondents with post-paid system. To be in line with the Javanese way of thinking, pre-paid system will allow them to observe and curb their monthly electricity consumption in comparison to the obtaining the total bill by the end of the month. This will promote their electrify consumption to the modest level. Hence, the pre-paid system should be the suggested solution to manage electricity consumption for Javanese people.

Meanwhile, in Bandung, the difference between post-paid and prepaid systems is significant. Respondents with post-paid system have higher awareness than respondents with prepaid system. However, in Bandung the prepaid system does not guide the people to put better awareness in electricity consumption than the post-paid system. The post paid system is very convenient for Sundanese people with their way of thinking in terms consumption; they consider their consumption not on daily basis, but in monthly. By simplifying in monthly consumption, then they can strategize to manage their electricity consumption for the upcoming month.

Furthermore, from the average frequency of respondents' response, electricity payment systems in Bandung and Yogyakarta serve as a signal in decision-making for electricity consumption. Together with results in Sect. 4.2, the monthly electricity consumption regardless the payment system is highly regarded by people to increase awareness of their consumption.

5 Conclusion

This study found that people in Bandung has significantly better understanding of electricity price as compared to people in Yogyakarta. If electricity prices were to be increased by 10% from current price, respondents in Yogyakarta had the tendency to reduce their electrical devices usage with significant differences (by statistical significance test less than 0.05) than in Bandung for TV, refrigerator, rice cooker, water pump, and washing machine, despite in the average frequency of respondents' response tend to not change their behavioral devices use. In terms of the effect of electricity payment system to monthly electricity use, respondents in Yogyakarta with prepaid system show slight better awareness in electricity payment (mean values are 4.23 for prepaid system and 4.14 for postpaid system), while for the respondents in Bandung it differs; respondents with post-paid system have higher awareness (mean values are 3.97 for prepaid system and 4.37 for postpaid system). Therefore, the effort to increase awareness on energy conservation should be to rectify the most proper electricity payment systems based on local characteristics in order to raise awareness of electricity usage. For future work, these results will be included and simulated quantitatively to measure the effectiveness of electricity payment system application based on local characteristics in the simulation of long-term electricity consumption.

Acknowledgment This work was partially supported by an "Energy Science in the Age of Global Warming" of Global Center of Excellence (G-COE) program (J-051) of the Ministry of Education, Culture, Sports, Science and Technology of Japan.

References

1. Ministry of Energy and Mineral Resources (2008) National Electricity Master Plan 2008–2027, Jakarta. Available at http://prokum.esdm.go.id/kepmen/2008/Kepmen%20ESDM%208202 682% 202008%20RUKN.pdf
2. Suryadunata L, Arifin EN, Ananta A (2003) Indonesia's population: ethnicity and religion in changing political landscape. Institute of Southeast Asian Studies, Singapore
3. Yamamoto Y, Suzuki A, Fuwa Y, Satu T (2008) Decision-making in electrical appliance use in the home. Energy Policy 36:1679–1686
4. Wijaya ME, Tezuka T (2012) Understanding socio-economic driving factors of Indonesian households electricity consumption in two urban areas. In: Yao T (ed) Zero-carbon energy Kyoto 2011. Springer, Tokyo, pp 56–60

Analysis of Intentions to Recycle Electronic Waste (E-Waste) Using the Theory of Planned Behavior: A Case Study in Urban Areas of Vietnam

Hoang-Long Le, Eiji Yamasue, Hideyuki Okumura, and Keiichi N. Ishihara

Abstract The theory of planned behavior (TPB) was used as a theoretical framework for systematically investigating the determinants of e-waste recycling behavior. The data for this study were collected from 1,806 households living in Hanoi city, Danang city and Hochiminh city, Vietnam. The findings suggest that TPB components explained a substantial proportion of variance in intentions to recycle e-waste (26.7%). The most important predictor was perceived behavioral control ($\beta = 0.331$, $p < 0.001$), the second strongest predictor was subjective norm ($\beta = 0.159$, $p < 0.001$) and the weakest was attitudes ($\beta = 0.108$, $p < 0.001$). Recycling habit and interest in economic benefit also were found to be significantly contributed to e-waste recycling intentions. Implication of the findings for establishing a management and recycling system specified for e-waste in Vietnam was discussed.

Keywords E-waste • Recycling intentions • The theory of planned behavior

1 Introduction

Electronic waste (e-waste) is commonly known as waste from electrical and electronic equipment (WEEE) or end-of-life (EOL) electronics [1]. It is the fastest growing segment of the municipal solid waste stream because of increased affordability of new products and technological achievements which make it easy to purchase of new electronics rather than repairing or upgrading old products [2]. There is no official statistic information about the current status of e-waste in Vietnam. However, it is predicted that the rate of e-waste will bloom very fast soon, and it will be a big problem if the is not good preparation for

H.-L. Le (✉) • E. Yamasue • H. Okumura • K.N. Ishihara
Graduate School of Energy Science, Kyoto University, Yoshida-honmachi, Sakyo-ku, Kyoto 606-8501, Japan
e-mail: lehoanglong.inest@gmail.com

T. Yao (ed.), *Zero-Carbon Energy Kyoto 2012*, Green Energy and Technology, DOI 10.1007/978-4-431-54264-3_7, © Springer Japan 2013

management capacity. Similar to other developing countries where specific regulations for management of e-waste are still not available, the collection, reuse, recycling and final disposal of e-waste are spontaneously driven by informal sectors. Such kind of system may lead to the serious environmental damages similar to what is happening in Guiyu—a large e-waste recycling site in China due to economic benefit is unique dynamic for operations. Facing with this challenge, Vietnamese government has announced a draft decision for collection and treatment of some special waste products including e-waste, which is planned to become effective from the beginning of 2014. According to this draft decision, e-waste will be managed following the extended producer responsibility principle, which has been applied in several developed countries. To implement effective policies to tackle the increasing stockpile of e-waste and to upgrade the existing recycling infrastructure, an understanding of households' willingness to recycle is necessary [3]. Thus, the aim of the study is to explore the determinants of residents' behavioral intentions to participate in e-waste recycling strategy according to the draft decision.

The theory of planned behavior (TPB) is a theory designed to predict and explain the human behavior in specific contexts [4]. The TPB emphasizes that human behaviors are governed not only by personal attitudes, but also by social pressures and a sense of control. According to this, the TPB hypothesizes that behavior is determined by intentions of individuals. Intentions are, in turn, influenced by three factors: (1) attitudes of individuals, the extent to which they perceive behavior as desirable or favorable, (2) the subjective norms, the degree to which significant individuals, such as relatives, friend, or colleagues, condone behavior [4], (3) perceived behavioral control (PBC), the extent to which individuals feel they can engage in these behaviors. In some special case the PBC also can impinge on behavior [4].

In general, human recycling behavior is determined by personal variables (internal motivators and facilitators) and situational variables (external motivators and facilitators) [5]. It looks simple, but there are many factors influencing human recycling behavior. The investigation of these factors requires a theoretical framework that basically covers both personal and situational variables, such as that provided by the TPB. In addition, in the past two decades, TPB has been applied in several studies which investigated recycling behavior with considerable success (see for example: [6–12]). For these notions, in this study the TPB incorporated a number of additional variables were used to investigate what affects the behavioral intentions to recycle e-waste in Vietnam. The finding of this study may lead to better understanding the attitudes of Vietnamese households with e-waste recycling. It also may be helpful for promoting effective e-waste recycling programs in Vietnam.

2 Method

2.1 Participants

The data for this study were collected in Hanoi city, Danang city and Hochiminh city, which are representative of three urban areas of three main regions in Vietnam. In each city, questionnaire based survey was organized by multistage cluster sampling method [13]. Questionnaires were distributed and collected by either door-to-door or at crowded sites where people have free time such as coffee shop and local administrative offices. The survey was completed between February and August 2011 with relatively good overall response rate (78.5%, N = 1806).

2.2 Questionnaire Design

Questionnaire was designed based on the recycling literatures and previous applications of TPB (see for example: [6, 9, 14]). The purpose of questionnaire is to gather information about the extent to which the respondents agree or disagree with a number of statements describing variables under investigation. The questionnaire contained the following sections:

- TPB's variables including intentions, attitude, subjective norm, perceived behavioral control. These variables are measured by 7-point Likert scales.
- Additional variables including new environmental paradigm (NEP), recycling habit, understanding about environmental laws and regularities, economic benefit, and environmental awareness. These variables are measured by 5-point Likert scales.
- Demographic information including age, gender, marital status, education level, ownership of house, house type, house size, monthly income.

3 Results

3.1 Descriptive Statistics and Correlations

Prior to analysis of TPB, descriptive statistics (means, standard deviations, and Cronbach's alpha coefficients) and bivariate correlations of all variables were calculated and the results were presented in Table 1. The results indicate that TPB's components had good internal consistency, with Cronbach's alpha coefficients reported from 0.80 to 0.95, whereas NEP had relatively low reliability

Table 1 Descriptive statistics and correlation between variables

Variables	N	M	SD	α	Intentions
Intentions	1,797	5.02	1.827	0.95	
Attitudes	1,797	6.06	0.985	0.86	0.237**
Subjective norm	1,797	5.37	1.403	0.80	0.384**
PBC	1,795	4.35	1.530	0.86	0.462**
NEP	1,794	3.63	0.411	0.62	0.034
Recycling habit	1,797	3.44	0.776	–	0.273**
Laws and regularities	1,797	3.17	0.747	–	0.213**
Economic benefit	1,795	3.46	0.785	–	0.246**
Environmental awareness	1,795	3.28	0.718	–	0.148**

**p < 0.01

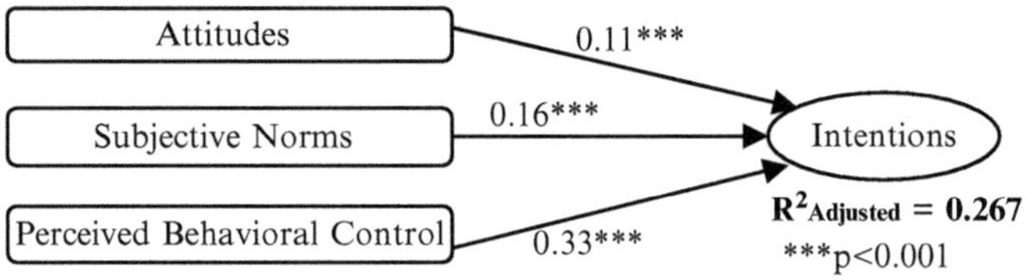

Fig. 1 Intentions to recycle e-waste (model 1)

with Cronbach's alpha coefficient of 0.62. All TPB's components were also significantly correlated with e-waste recycling intentions. Significant correlations were also found between intentions and all additional variables, except NEP.

3.2 Multiple Regression Analysis

Hierarchical multiple regression was used to investigate the determinants of e-waste recycling behavior. At first model, the components of TPB (attitudes, subjective norm, and PBC) were entered. As shown in Fig. 1, these three components collectively explained 26.7% of the variance in e-waste recycling intentions, with all three components being statistically significant predictors at 0.1% level [$F_{Change}(3,1787) = 218.048$, $p < 0.001$].

In the model 2 (Fig. 2), the additional variables including recycling habit, environmental laws and regularities understanding, interest in economic benefit, and environmental awareness were entered the multiple regression in succession. NEP was excluded from the regression analysis because of its low reliability and its insignificant correlation with dependent variable. The analysis results show that recycling habit significantly contributed an additional 1.1% to the variance of the outcome [$R^2_{Change} = 0.011, F_{Change}(1,1786) = 26.403, p < 0.001$]. Interest in economic benefit significantly contributed further 0.8% to the explanation of variance [$R^2_{Change} = 0.008, F_{Change}(1,1784) = 20.591, p < 0.001$]. Environmental laws and

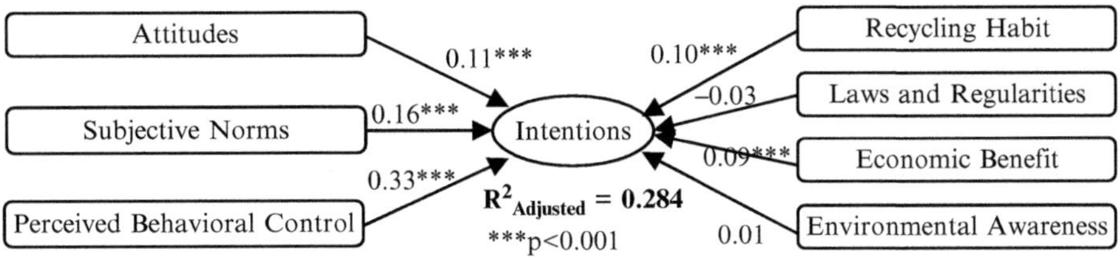

Fig. 2 Intentions to recycle e-waste (model 2)

regularities understanding [$R^2_{Change} = 0.000, F_{Change}(1{,}1785) = 0.069, p = 0.793$] and environmental awareness [$R^2_{Change} = 0.000$, $F_{Change}(1{,}1783) = 0.332$, $p = 0.564$] were found to be non-significantly contributed to the variance.

In summary, the total variance explained by the model 2 as a whole was 28.4% [$R^2_{Adjusted} = 2.84$, $F(7{,}1783) = 102.540$, $p < 0.001$], with PBC making the strongest unique contribution to explaining the intentions to recycle e-waste ($\beta = 0.33$, $p < 0.001$), and followings were subjective norm ($\beta = 0.16$, $p < 0.001$), attitudes ($\beta = 0.11$, $p < 0.001$), interest in economic benefit ($\beta = 0.10$, $p < 0.001$), and recycling habit ($\beta = 0.09$, $p < 0.001$).

4 Discussion

As expected, the application of TPB as a theoretical framework for systematically investigating psychological factors influencing e-waste recycling behavior in Vietnam is highly informative. The original TPB model explained totally 26.7% of the variance in intentions to recycle e-waste. All attitudes, subjective norm and PBC were significantly contributed to recycling intentions in the expected directions. The variance explained was somewhat less than the average value of 39% reported by Armitage and Conner [15]. However, the result of this study is supported by Cheung et al. [11] who argue that in fact, TPB alone could explain 27.7–29.1% of the variance in self-reported behavior. And more recent studies also reported similar result, Tonglet et al. [8] reported the TPB only explained 21.6% of the variance in recycling intentions, and Knussen et al. [9] reported that value of 21.6% and 29%, respectively.

In this study, when additional variables were included into the model, the percentage of variance explained increased from 26.7% to 28.4%. Although, these additional variables only contributed additionally small portion of variance explained, the statistic significance of this finding supports the inclusion of additional variables when apply the TPB to recycling behavior as Boldero [12] and Davies et al. [7] concluded. Of the additional variables investigated, recycling habit and interest in economic benefit were significantly contributed to the explanation of variance in intentions to recycle e-waste. The similar result also was indicated in Wang et al. [2].

The subjective norm variable was the second strongest predictor of recycling intentions. This result was different from most previous TPB studies. Terry et al.

[16], Armigate and Conner [15] noted that subjective norm is often more weakly related to intentions than are the other TPB components. In Davis et al. [7] and Tonglet et al. [6] the subjective norm was the weakest predictor of the TPB variable. In Knussen et al. [9] and Boldero [12] the subjective norm was not significant at all. However, the result in this study could be explained by the fact that Vietnamese households treat electronics as a family's property even they are obsolete or unwanted because they still have a considerable value. Thus, the e-waste disposal behavior is not only dependent on an individual but also is strongly affected by family members, friends or neighbors.

With respect to the perceived behavioral control measure, this study operationalized perceived behavioral control by mixing of traditional variables (easy and opportunity) and facilitating or inhibiting factors (knowledge of what, how and where to recycle). The significance contribution of perceived behavioral control as a predictor of recycling intentions not only supports the suggestion that situational factors are likely to be important to recycling behavior [12] but also consists with the notation that control factors which facilitate or inhibit the performance of the behavior in question provide a more accurate measure of PBC than the measure normally used [7].

5 Conclusions

The application of the theory of planned behavior (TPB) in determining psychological factors influencing e-waste recycling behavior in Vietnam was developed. The results indicated that TPB components explained a moderate proportion of variance in intentions to recycle e-waste. The most important predictor was PBC, the second strongest predictor was subjective norm and the weakest was attitudes. The significant contribution and positive prediction direction of all TPB components in this study support TPB hypothesis.

Hierarchical multiple regression analysis shown the significant increase of variance explained, which supports the inclusion of additional variables when apply the TPB to recycling behavior. However, the additional variables investigated can only contribute a small additional portion of variance explained. Therefore developing other additional variables is needed in future studies.

The findings from this study have several implications for the establishment of management and recycling system specified for e-waste in Vietnam.

- Since perceived behavioral control was the strongest predictor of e-waste recycling intentions, top priority should be given to the construction of e-waste recycling infrastructure.
- Given that attitudes and subjective norm also were significant predictors of e-waste recycling intentions, communication campaigns should be designed to reinforce the positive attitudes as well as change the negative attitudes of people by emphasizing both effective (feelings about recycling) and cognitive

(knowledge of the outcomes and consequences) aspects of e-waste recycling. In addition, an effective monitoring system at community level creating sufficient social pressure on e-waste recycling also very necessary.

- Finally, the findings from this study note that recycling habit also significantly contributed to e-waste recycling intentions. This implies that recycling habit should be promulgated widely and deeply by educational programs from the child.

Acknowledgment This work was partly supported by an "Energy Science in the Age of Global Warming" of Global Center of Excellence (G-COE) program (J-051) of the Ministry of Education, Culture, Sports, Science and Technology of Japan.

References

1. Cairn CN (2005) E-waste and the consumer: improving options to reduce, reuse and recycle. In: Proceedings of the 2005 I.E. international symposium on electronics and the environment, New Orleans, LA, USA, pp 237–242
2. Sawhney P, Henzler M, Melnitzky S, Lung A (2008) Best practices for E-waste management in developed countries. Adelphi Research, Berlin
3. Saphores JDM, Nixon H, Ogunseitan OA, Shapiro AA (2006) Household willingness to recycle electronic waste: an application to California. Environ Behav 3–2:183–208
4. Ajen I (1991) The theory of planned behavior. Organ Behav Hum Decis Processes 50:179–211
5. Hornick J, Cherian J, Mandansky M, Narayna C (1995) Determinants of recycling behavior: a synthesis of research results. J Socioecon 24:105–127
6. Tonglet M, Phillips PS, Read AD (2004) Using the theory of planned behavior to investigate the determinants of recycling behavior: a case study from Brixworth, UK. Resour Conserv Recycl 41:191–214
7. Davis G, Phillips PS, Read AD, Iida Y (2006) Demonstrating the need for the development of internal research capacity: understanding recycling participation using the theory of planned behavior in West Oxfordshire, UK. Resour Conserv Recycl 46:115–127
8. Chu PY, Chiu JF (2003) Factors influencing household waste recycling behavior: test of an integrated model. J Appl Soc Psychol 33(3):604–626
9. Knussen C, Yule F, MacKenzie J, Wells M (2004) An analysis of intentions to recycle household waste: the roles of past behavior, perceived habit, and perceived lack of facilities. J Environ Psychol 24:237–246
10. Taylor S, Todd P (1995) An integrated model of waste management behavior: a test of household recycling and composting inventions. Environ Behav 27:603–630
11. Cheung SF, Chan DKS, Wong ZSY (1999) Reexamining the theory of planned behavior in understanding wastepaper recycling. Environ Behav 31:587–612
12. Boldero J (1995) The prediction of household recycling of newspaper: the role of attitudes, intentions and situational factors. J Appl Soc Psychol 25(5):440–462
13. Buckingham A, Saunders P (2004) The survey methods workbook. Polity, Cambridge, pp 119–121
14. Wang Z, Zhang B, Yin J, Zhang X (2011) Willingness and behavior towards e-waste recycling for residents in Beijing city, China. J Clean Prod 19:984–997
15. Armitage CJ, Conner M (2001) Efficacy of the theory of planned behaviour: a meta-analytic review. Br J Soc Psychol 40:471–499
16. Terry DJ, Hogg MA, White KM (1999) The theory of planned behavior: self-identity and group norms. Br J Soc Psychol 38:225–244

A Design Method of Online Community for Behavior Change Focusing on Participants' Relationship

Saizo Aoyagi, Hirotake Ishii, and Hiroshi Shimoda

Abstract Behavior change is sometimes needed to solve energy problems. A design method of online community for behavior change, which is a socio-psychology based systematic method, was proposed. The features of the proposed methods are (1) adaptation of community to behavior change stage of participants and social recommendation level of target behavior and (2) designing online community with communication content and strength of relationship between members.

Keywords Behavior change • Online communication • Online community • Relationship

1 Introduction

Human behaviors are deeply related to processes of energy use. For example, human operate, maintain and repair power plants and supply line of energy. In addition, human utilize many electric appliances. Therefore behavior change of some people is often demanded for safety or efficient energy use, such as promotion of energy saving behavior in house and office, or promotion of safety behavior in works in power plants.

This study focused on utilization of online community as a useful method for behavior change. Mutual support, knowledge exchange, group norm or other socio-psychological powers help participants continue to effort for behavior change. Additionally, asynchronous and distributed online communication prevents privacy problems and saves time for setting face-to-face meeting for community.

S. Aoyagi (✉) • H. Ishii • H. Shimoda
Graduate School of Energy Science, Kyoto University, Yoshida-honmachi, Sakyo-ku Kyoto, Kyoto 606-8501, Japan
e-mail: aoyagi@ei.energy.kyoto-u.ac.jp

T. Yao (ed.), *Zero-Carbon Energy Kyoto 2012*, Green Energy and Technology,
DOI 10.1007/978-4-431-54264-3_8, © Springer Japan 2013

On the other hand, online communities for behavior change often cause some problems; participants drop out from communities or activities of communities are sometimes not effective for participants' behavior change. Therefore, design method of online community for behavior change in order to avoid problems and make effective online community for behavior change.

There exist comparatively few attempts to build systematic online communities design method [1], because dominating methodology for online community design is best practice and pattern approach. These retrospective approaches however lacked general recipes for making online community from scratch [2]. In addition, very few study of online community design deal with its effects on behavior in the real world of participants, because main concerns of these studies are behaviors in online world.

A purpose of this paper is proposal of a design method of online community for behavior change. The proposed method provides one of the few systematic ways of online community design, which is based on social psychological theories, and is designed to activate not only participants' behaviors in online world but also behaviors in online world. It will be a useful tool for managers of online community for various kinds of behavior change.

2 Proposal of a Design Method for Behavior Change

2.1 Principles of the Method

First, the proposed method presumes text-based, asynchronous and distributed online communication. Participants freely join activities of online community whenever they choose using information devices such as smartphone. This type of free participation is very convenient for participants. Nevertheless, it possibly causes dropout of participants.

Therefore, online community for behavior change has to activate both participants' behaviors in online world and these in real world. Activation of target behavior in real world is a purpose of online community for behavior change. Additionally, activation of behaviors in online world means activation of communication in online community, which is needed for causing socio-psychological powers and keeping participant in activities of online community to avoid dropouts. Designing for meeting these two acquirements, activation of online and real behaviors, is a basic principle of the proposed method.

In addition, there is another principle. According to behavior change theories, suitable intervention is depends on target people, target behavior, and situations surrounding them [3]. Therefore, online communities for behavior change have to adapt or suitably designed to participants, target behavior, and situations.

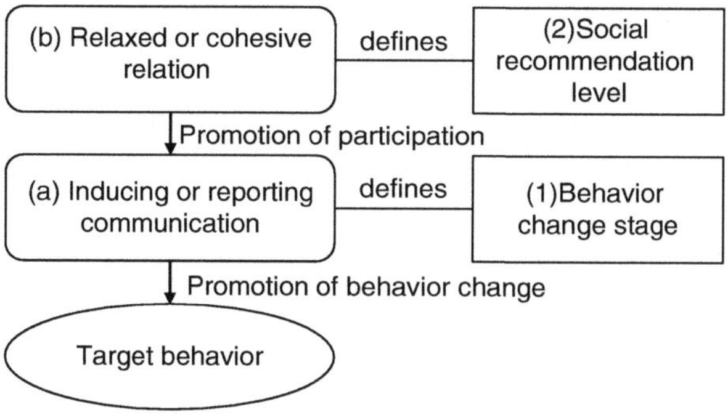

Fig. 1 A framework of the design method

2.2 Stage of Behavior Change and Communication Content

Trans-theoretical model (TTM) [3] pointed out that behavior change process generally consist of five stages. Stages in this context mean psychological conditions related to behavior change. Each stage is distinguished from others by presence or absence of some kinds of intention, cognition or experience. In addition to that, people in each stage need different intervention in order to move to next stage.

For example, people in "preparation stage" have intention to do target behavior; they however do not do that yet. They need to feel that "target behavior is available" and declaration of doing target behavior to the others in order to move to "action stage". On the other hand, people in "action stage" have an experience of doing target behavior; they however do not continue that. They need incentive control or need to know that others do target behavior in order to move to "maintenance stage".

As mentioned in Sect. 2.1, online community for behavior change has to activate promote behaviors in real world, and TTM points out that effective intervention for activation of real world behavior depends on participants' stage of behavior change. Therefore, the proposed design method utilizes stage of TTM, and makes online community adapt to participants in each stage, and provide suitable intervention for each of them.

In concrete terms, the design method adapts contents of communication in activities of online community to stage of behavior change of participants. Figure 1 shows a framework of the proposed method, and bottom parts of this figure show that suitable (a) communication content to promote target behavior is defined by (1) behavior change stage of participants.

People in preparation stage do "inducing communication". In inducing communication, "a supporting element", such as norm induces doing target behavior. Next, participants who did target behavior report the fact that they did that to the other behavior. At last, "feeling of being looked element", such as replies for these reports enhances their feeling that behaviors are looked by other participants.

Inducing communication implement declaration and getting self-confidence, and helps people in action stage move to next stage.

People in action stage do "reporting communication". In this kind of communication, participants report the fact that they did target behavior each other and continuously. This mutual reporting will affect participants, causes conformation effect, participants will be able to continue target behavior and move to "maintenance stage".

2.3 Recommendation Level and Relation of Participants

Another requirement for online community for behavior change is activation of behaviors (communication) in online world, in order to avoid dropout from the activity of online community as mentioned in Sect. 2.1.

The authors propose relationship among participants of online community as a tool for meeting this requirement. This is because relationships are unavoidably formed in every community and it strongly effect participants' motivation to participate to communication in online community.

The authors proposed two kinds of relations whose strength are different, and each of them is suitable for different social recommendation levels. Social recommendation level is a concept that declares a degree of recommendation of target behavior in organization or society which participants belong to.

"Relaxed relation" is affable, moderate form of human relationship. It resembles just sharing a place. Group norm and social influences are not strong. It is easy to join and suitable for participants in low motivation participants, who are not in charge of target behavior in their organization or society. This state is defined as "low recommendation level".

"Cohesive relation" is strong, tight form of human relationship. Roles and group norm definitely exist and social influences are strong. It is suitable for high motivation participants who are in charge of target behavior, and this stage is defined as "high recommendation level".

A top parts of Fig. 1 shows that (2) social recommendation level of target behavior defines (b) suitable type of relation in online community, and relation will promote participation of online community, which means behavior in online world.

Figure 2 summarizes the proposed method and its design procedure. First, designer of online community for behavior change determines stage of behavior change of participants and social recommendation level of target behavior. Next, s/he determines (a) communication content and (b) strength of relation. At last, develop information system for online community in order to embody communication content and strength of relation.

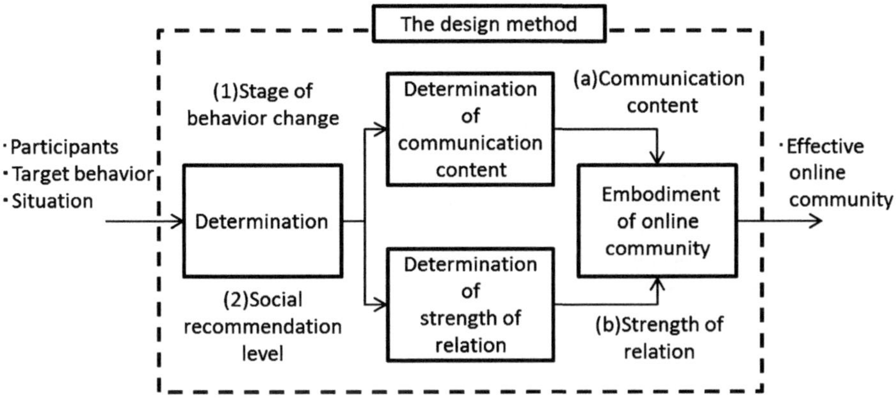

Fig. 2 The outline and its design procedure of the design method

3 Conclusion

The design method for online community for behavior change was proposed. This is one of the few systematic ways of online community design, and is designed to activate not only participants' behaviors in online world but also behaviors in online world. In addition, concepts of strength of relationship has original point of view of this study. There are many studies that deal with bond or relation between humans [2], relaxed relation was not so focused as useful tool for behavior change.

Because of limitation of paper, details of the method and procedure of information system development were not described. Moreover, evaluation of the proposed method was not described in this paper. In fact, this design method was applied to three behavior change problems and designed communities were evaluated through experiments with participants [4–6], and results of experiments of the method showed that it is useful for design of online communities that are effective for behavior change.

It will be a useful tool for managers of online community for various kinds of behavior change.

Acknowledgement The authors were grateful for the support of the Ministry of Education, Culture, Sports, Science and Technology of Japan via "Energy Science in the Age of Global Warming" of Global Center of Excellence (G-COE) program (J-051).

References

1. Ren R, Kraut R, Kiesler S (2007) Applying common identity and bond theory to design of online communities. Organ Stud 28:377–408
2. Vassileva J (2012) Motivating participation in social computing applications: a user modeling perspective. User Model User Adap Inter 22:177–201

3. Prochaska JO (1979) Systems of psychotherapy: a transtheoretical analysis. Dorsey, Homewood
4. Aoyagi S, Fujino H, Ishii H, Shimoda H, Sakura H, Yoshikawa H, Sugiman T (2011) Proposal and field practice of a "Hiyarihatto" activity method for promotion of statements of participants for nuclear power plant organization. Trans Atomic Energy Soc Jpn 10–4:273–289
5. Aoyagi S, Fujiwara O, Ishii H, Shimoda H (2011) Application of online community to promotion of pro-environmental behavior. In: The 3rd international symposium: Kyoto University Global COE Program "Energy Science in the Age of Global Warming—Toward CO2 Zero-emission-"(Zero-Carbon Energy Kyoto 2011)
6. Aoyagi S, Okamura T, Ishii H, Shimoda H (2011) Proposal and evaluation of a method for promoting continuous pro-environmental behavior with moderate communication. Pap Hum Interface Soc 13–3:31–44

Part II
Renewable Energy Research and CO_2 Reduction Research

(i)
Session Papers

Effect of Water and Free Fatty Acids in Oil on Biodiesel Production by Supercritical Methyl Acetate Method*

Fadjar Goembira and Shiro Saka

Abstract Some oils for biodiesel contain high water and free fatty acids (FFA) that limits their use for biodiesel production by acid-catalyzed and alkali-catalyzed methods. This study was, thus, conducted to know the effect of those compounds on supercritical methyl acetate method, as an alternative to the catalyzed methods. Particularly at the shorter reaction times, water and FFA increased the reaction rate in the non-catalytic supercritical methyl acetate method, thus, the higher yields were found to be evident, as compared to these catalyzed methods.

Keywords Biodiesel • Fatty acid methyl ester • Free fatty acid • Supercritical methyl acetate • Water

1 Introduction

The increase of biodiesel production worldwide has caused an excessive use of refined plant oils. This situation is due to the unavoidable use of feedstocks with low water and free fatty acid (FFA) contents for the conventional acid-catalyzed and alkali-catalyzed transesterification, which are the most popular biodiesel production methods at the moment.

In order to attain feedstocks with low water and FFA contents, refined edible oils from rapeseed (*Brassica napus*) and soybean (*Glycine max*) are currently used for biodiesel production in Europe and the United States, respectively. From year 1999–2006, as the world biodiesel production significantly increased, the prices

*This work is an excerpt of PhD dissertation by F. Goembira in Kyoto University who is presently in Department of Environmental Engineering, Faculty of Engineering, Andalas University, Indonesia.

F. Goembira • S. Saka (✉)
Graduate School of Energy Science, Kyoto University, Kyoto, Japan
e-mail: saka@energy.kyoto-u.ac.jp

T. Yao (ed.), *Zero-Carbon Energy Kyoto 2012*, Green Energy and Technology, DOI 10.1007/978-4-431-54264-3_9, © Springer Japan 2013

of rapeseed and soybean oils were also considerably increased from a similar price of around 350 USD/tonne to 850 and 650 USD/tonne, correspondingly [1]. Direct consequences of this condition will be higher biodiesel production cost and an inevitable increase of food price.

In addition, supercritical methyl acetate method has been developed to produce fatty acid methyl esters (FAME) and triacetin, as a by-product, through an interesterification reaction [2]. FAME is miscible with triacetin and the mixture was found to improve some biodiesel properties, e.g., cold-flow properties and oxidation stability [3–5], thus, this mixture can potentially be used as biodiesel.

This study was conducted to know the effect of water and FFA in plant oils on biodiesel production by supercritical methyl acetate method. For a comparison, the effects of those compounds on the acid-catalyzed and alkali-catalyzed methods were also studied. Based on the results, we expect to know the possible utilization of the supercritical process on plant oils with high water and FFA contents.

2 Experimental

Model oils were prepared to simulate different water and FFA contents that present in biodiesel feedstocks, by mixing various amounts of triglycerides (rapeseed oil), water and oleic acid as FFA as shown in Table 1. The use of high water and FFA contents in the model oils, i.e., up to 5 and 30 wt.%, respectively, was to find out the applicability of the three biodiesel production methods on some potential feedstocks that have relatively high water and FFA contents, such as waste palm oil [6].

Moreover, high purity (>99%) methyl acetate, methanol, analytical grade FAME and triacetin as standard compounds were used. In addition, sulfuric acid and sodium hydroxide were utilized as the catalysts for the acid-catalyzed and alkali-catalyzed methods, respectively. All compounds were bought from Nacalai Tesque, Inc. Kyoto, Japan.

Acid-catalyzed method was conducted by using a molar ratio of 30:1 in methanol to oil, with 3 wt.% catalyst, under 48 h reaction time at 65°C [7]. In the alkali-catalyzed method, transesterification was proceeded under 6:1 M ratio of methanol to oil, with 1.5 wt.% catalyst at 60°C for 1 h [8]. Whereas for the supercritical methyl acetate method, interesterification was carried out at 42:1 M ratio at 350°C/20 MPa for 45 min [9]. The difference in solvent to oil molar ratio was necessary in order to obtain the highest yield, as demonstrated by previous studies [7–9]. In addition, for the supercritical method, the shorter reaction time of 10, 15 and 30 min was also used in this study.

Table 1 The model oils used in this study that contain various amounts of water and oleic acid as FFA

Component	Chemical composition (wt.%)															
Triglycerides	100	99	97	95	95	90	70	94	89	69	92	87	67	90	85	65
Water	0	1	3	5	0	0	0	1	1	1	3	3	3	5	5	5
Oleic acid	0	0	0	0	5	10	30	5	10	30	5	10	30	5	10	30

Round-bottom flask, condenser and mantle heater were used in the catalyzed methods, while a batch-type reaction vessel with gas-charging inlet, thermocouple and pressure gauge was utilized for the supercritical method. The obtained products were analyzed by a high performance liquid chromatography (HPLC) with Cadenza CD-C18 as the column under 40°C HPLC oven temperature. Methanol was used as the mobile phase in the HPLC analysis.

3 Results and Discussion

3.1 Effect of Water and FFA on FAME Yield

Figure 1 shows the effect of water and FFA in oils on FAME yields from (a) acid-catalyzed, (b) alkali-catalyzed and (c) supercritical methyl acetate methods.

In the acid-catalyzed method, higher water or FFA in oils resulted in the lower yields. For instance, the yields were dropped from 98 wt.% when those compounds were not present, to 80 or 90 wt.% when the oils had 1 wt.% water or 5 wt.% FFA, respectively. Additionally, yield reduction to 65 wt.% was evident when both 1 wt.% water and 5 wt.% FFA were present in the oil. Yet, observable yield reduction when no water was present in oil might be due to the esterification reaction between methanol and FFA that produces FAME and water in the reaction system.

Moreover, as shown by Fig. 1b, in the alkali-catalyzed method, more water or FFA in oils also resulted in the lower FAME yields. In the absence of those compounds, however, the yield was 97 wt.%. Nevertheless, when the oils had 1 wt.% water or 5 wt.% FFA, the yields were reduced to 90 or 60 wt.%, accordingly. Furthermore, a worse yield reduction to 50 wt.% happened due to the simultaneous presence of 1 wt.% water and 5 wt.% FFA in oil.

Interestingly, in the supercritical methyl acetate method as shown by Fig. 1c, reductions of FAME yields were not as significant as in these two catalyzed methods. For example, FAME yields when the oils contained 1 wt.% water or 5 wt.% FFA were 92 or 96 wt.%, respectively, as compared to the 97 wt.% yield without the presence of these compounds. Insignificant yield reduction to 90 wt.% was also noticeable when the oil contained 1 and 5 wt.% water and FFA, correspondingly.

Water gave more impact than FFA to the acid-catalyzed and supercritical methyl acetate methods, although insignificant impact was observable in the latter one. On the other hand, water might hydrolyzed some formed products in the supercritical method, thus yield reductions might occurred [6]. Meanwhile, in the alkali-catalyzed method, FFA gave more detrimental effect than water on FAME yields, because FFA reacts with alkali catalyst to produce saponified products. Besides the reduction of reaction rate due to catalyst consumption, lower yields were also likely due to the difficulties in separating FAME with glycerol because of the formation of saponified products [8]. In all cases, reductions of FAME yield were found to be more evident under the simultaneous presence of water and FFA in oils.

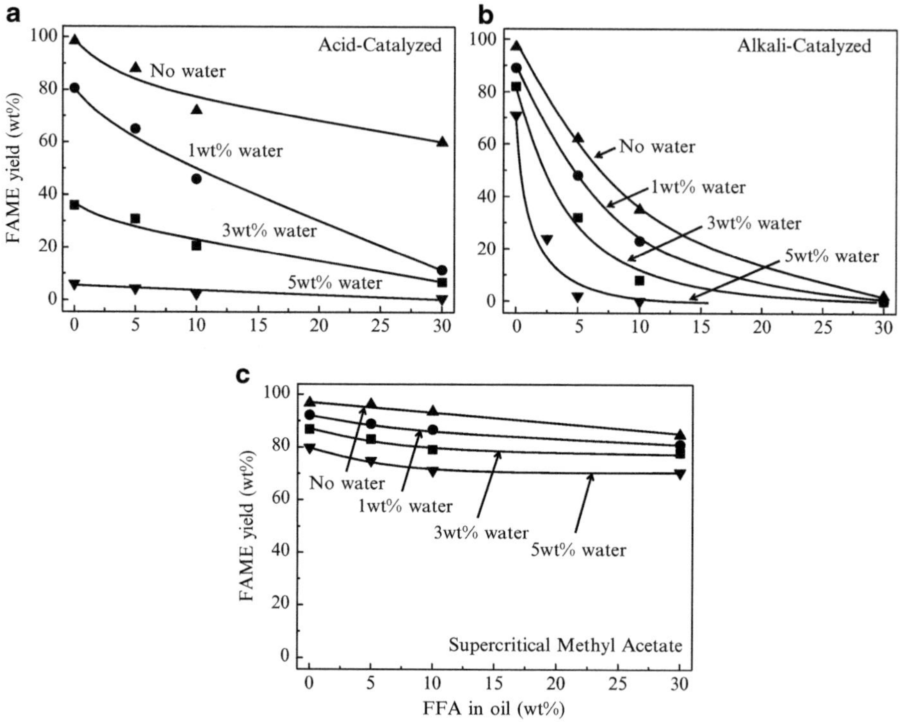

Fig. 1 FAME yields from model oils as treated in (**a**) acid-catalyzed, (**b**) alkali-catalyzed and (**c**) supercritical methyl acetate methods

3.2 *Roles of Water and FFA in Plant Oils on Transesterification/Interesterification*

We can see the results of the transesterification/interesterification reactions from HPLC chromatograms as shown in Fig. 2. In the acid-catalyzed and alkali-catalyzed methods, the peaks of intermediates are very observable, while in the supercritical methyl acetate method, only the peaks of FAME and triacetin are exhibited. It can, therefore, be concluded that the presence of water and FFA in oils only inhibits transesterification reaction in the catalyzed methods.

Since no intermediate peaks are observable, with the presence of water and FFA decreasing FAME yields in the non-catalytic supercritical method, it is interesting to evaluate the yields under the shorter reaction times. As shown in Fig. 3, the presence of water and FFA has increased the yields particularly at shorter reaction times. At 10 min reaction time, for example, the absence of water and FFA resulted in 30 wt.% yield; however, the presence of 30 wt.% FFA increased the yield to 45 wt.% and when there were 5 wt.% water and 30 wt.% FFA, the yield was raised to 48 wt.%. Water presumably hydrolyzed the triglycerides to produce more FFA in

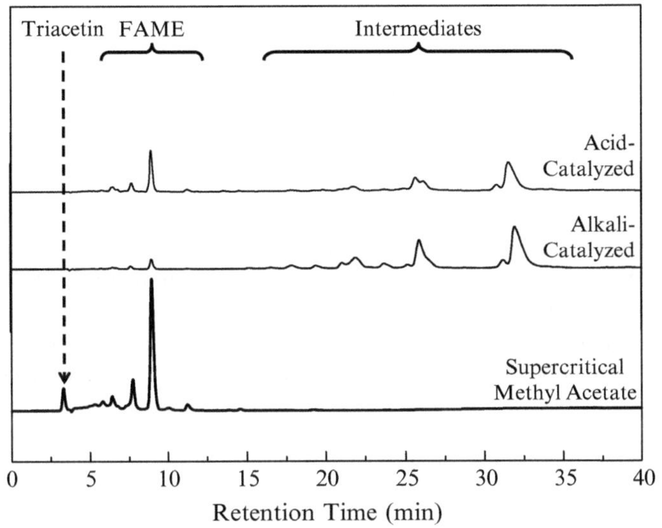

Fig. 2 HPLC chromatograms of biodiesel obtained from oil with 3 wt.% water and 10 wt.% FFA

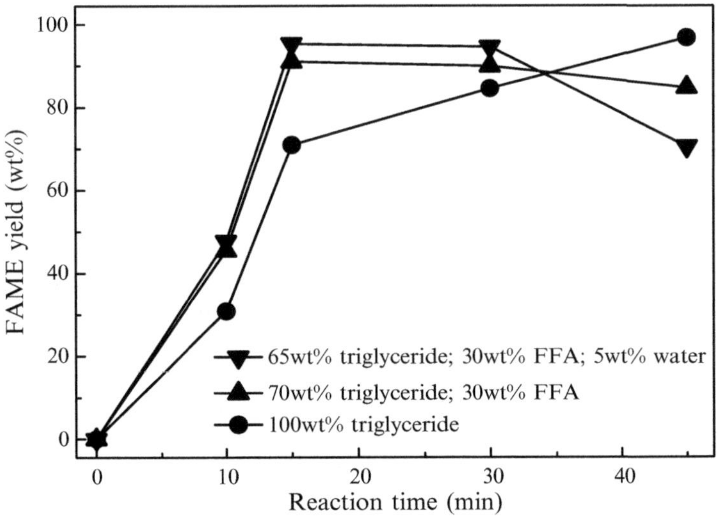

Fig. 3 Effect of water and FFA in oils on FAME yield at various reaction times by supercritical methyl acetate method

the reaction system. FFA has higher reactivity towards supercritical methyl acetate [3], thus, additional FFA due to the hydrolyzed triglycerides could generate more FAME. However, from the figure, we can see that the higher yields due to the presence of water and FFA were only up to 30 min reaction time, reducing the yield afterwards due perhaps to the degradation of FAME under supercritical conditions.

4 Conclusions

This study has demonstrated that the simultaneous presence of water and FFA in oils reduced the yield in biodiesel production by acid-catalyzed, alkali-catalyzed and non-catalyzed supercritical methyl acetate methods. However, different roles of water and FFA on the catalyzed and the supercritical methods were discovered. In the catalyzed methods, water and FFA reduced the reaction rate of transesterification, thus, a significant yield reduction was unavoidable. Conversely, water and FFA have increased the reaction rate in supercritical methyl acetate method, particularly at shorter reaction times. It can, therefore, be concluded that supercritical methyl acetate method is not vulnerable to the presence of water and FFA, which makes it as a potential process for broader ranges of biodiesel feedstocks.

Acknowledgement This study was completed under financial support from the Global-COE Program "Energy Science in the Age of Global Warming", Kyoto University, for all of which the authors highly acknowledge.

References

1. Durret TP, Benning C, Ohlrogge J (2008) Plant triacylglycerols as feedstocks for the production of biofuels. Plant J 54:593–607
2. Saka S (2005) Manufacturing method for fatty acid methyl esters. Japan Patent 4,378,534 (application: 19 Dec 2005; granted: 2 Oct 2009)
3. Saka S, Isayama Y (2009) A new process for catalyst-free production of biodiesel using supercritical methyl acetate. Fuel 88:1307–1313
4. Goembira F, Matsuura K, Saka S (2012) Biodiesel production from rapeseed oil by various supercritical carboxylate esters. Fuel 97:373–378
5. Casas A, Ruiz JR, Ramos MJ, Perez A (2010) Effects of triacetin on biodiesel quality. Energy Fuel 24:4481–4489
6. Kusdiana D, Saka S (2004) Effects of water on biodiesel fuel production by supercritical methanol treatment. Bioresour Technol 91:289–295
7. Canakci M, Van Gerpen J (1999) Biodiesel production via acid catalysis. Trans ASAE 42:1203–1210
8. Ma F, Clements LD, Hanna MA (1998) The effects of catalyst, free fatty acids and water on transesterification of beef tallow. Trans ASAE 41:1261–1264
9. Goembira F, Saka S (2012) Factors affecting biodiesel yield in interesterification of rapeseed oil by supercritical methyl acetate. In: Yao T (ed) Zero-carbon energy Kyoto 2011, green energy and technology. Springer, Tokyo, pp 147–152

Reactivity of Triglycerides and Fatty Acids in Sub/Supercritical Dialkyl Carbonates for Biodiesel Production*

Zul Ilham and Shiro Saka

Abstract Recently, interest in dialkyl carbonates as an environmentally benign chemical for green chemistry is increasing but its utilization for biodiesel production in non-catalytic supercritical method has never been discussed earlier. Therefore, we studied the reactivity of triglycerides and fatty acids with several dialkyl carbonates in the non-catalytic sub/supercritical treatment in order to evaluate the potential of various dialkyl carbonates for the production of fatty acid alkyl esters (FAAE) and other value-added by-products. In batch-type and flow-type reaction systems, reaction condition was set at 300°C/20 MPa when dimethyl carbonate, diethyl carbonate, dipropyl carbonate and dibutyl carbonate have been utilized as reactants. The results showed that esterification of fatty acids proceeds at higher reaction rate than the interesterification of triglycerides in non-catalytic supercritical dialkyl carbonates. Shorter alkyl chains of dialkyl carbonates also showed higher reactivity than the longer ones. Based on the obtained results, potential of dialkyl carbonates for biodiesel production in non-catalytic supercritical process was discussed.

Keywords Biodiesel • Dialkyl carbonates • Fatty acid alkyl esters • Interesterification • Non-catalytic supercritical method

---

*This work is an excerpt of PhD dissertation by Z. Ilham in Kyoto University who is presently in Institute of Biological Science, University of Malaya, Kuala Lumpur 50603, Malaysia.

Z. Ilham • S. Saka (✉)
Graduate School of Energy Science, Kyoto University, Kyoto 606-8501, Japan
e-mail: saka@energy.kyoto-u.ac.jp

T. Yao (ed.), *Zero-Carbon Energy Kyoto 2012*, Green Energy and Technology,
DOI 10.1007/978-4-431-54264-3_10, © Springer Japan 2013

1 Introduction

The world is currently at the edge of transition from totally relying on fossil fuel towards a bio-based economy in which, biomass resources hold the key. Among all, interesterification of oils and fats from biomass resources with short-chained alcohols and alkali catalyst to produce biodiesel or fatty acid alkyl esters (FAAE) and glycerol as a by-product has been common in the past decades [1].

It is known that crude glycerol generated from the conventional alkali-catalyzed process for biodiesel production contains a significant amount of impurities such as alcohols, salts, heavy metals and water [2]. It is therefore economically and environmentally questionable to process this extremely low value and high-volume mixtures of impurities into pure glycerol, which is readily available in large quantity. This crude biodiesel glycerol needs further purifications in order to produce pure glycerol, and the cost will be multiple times higher than producing glycerol in the conventional way [1].

Due to these issues, it is better if biodiesel production produces no glycerol as a by-product. Some possible routes are through interesterification of triglycerides with supercritical methyl acetate [3]. Another glycerol-free process for biodiesel production using non-catalytic subcritical acetic acid and supercritical methanol has also been introduced [4].

In addition to those, dimethyl carbonate, a versatile environmentally-friendly compound obtained from methanol and carbon dioxide is also a very promising reactant to produce biodiesel with non-glycerol high-value by-products [5]. Several earlier publications have also described this potential either in a non-catalytic [6, 7], catalytic or enzymatic production routes. However, utilization of other dialkyl carbonates for biodiesel production in non-catalytic supercritical process has never been studied earlier.

In this study, therefore, the reactivity of triglycerides and fatty acids in various dialkyl carbonates have been studied in sub/supercritical condition without any catalyst applied. This information is useful to charter potentially new pathways for biodiesel production without producing glycerol.

2 Experimental

Triglycerides from rapeseed oil, fatty acids, various authentic compounds for standards and chemicals were obtained from Nacalai Tesque Inc., Japan, all of which are of highest purity available.

Experiments were carried out in the flow-type reaction system with tubular reaction vessel made from Hastelloy C-276, where temperature and pressure could be constantly controlled and the flow capacity of this system is up to 600 mL/h. Triglycerides and various dialkyl carbonates were pumped into the system from different tanks and the molar ratio was determined by varying the pump flow-rate. Reactions employed in this study were in the range between 270–400°C and 10–40 MPa, without any catalyst applied. All reactions were monitored and controlled by thermocouple and pressure gauge. Initially, bath-type reaction system (5 mL Inconel-625 reaction vessel) was also used in this study. Details for both systems have been described in previous works [4, 6, 7]. Products were, then, analyzed by gel permeation chromatography (GPC) (column: GF-310HQ, column temp.: 40°C, flow rate: 1 mL/min, mobile phase: acetone, detector: RID 10A) and high performance liquid chromatography (HPLC) (column: Cadenza CD-C18, column temp.: 40°C, flow rate: 1 mL/min, mobile phase: methanol, detector: RID 10A). Glycerol carbonate was analyzed by HPLC (column: Ultrahydrogel 120, column temp.: 40°C, flow rate: 1 mL/min, mobile phase: water, detector: RID 10A). Gasified products whenever applicable were studied with Micro Gas Chromatograph CP 4900 (column: PoraPLOT Q 10 m, column temp.: 80°C, carrier gas pressure: 190 kPa, carrier gas: Argon and Helium).

Confirmation of the obtained products was identified by comparison of retention times with those of authentic reference compounds. For this purpose, the calibration curve of the authentic standard products, triolein, monoolein, diolein, glycerol carbonate and methyl oleate were used. All sets of experiments were made to at least triplicate for confirmation of the yields, although not treated statistically. The yield of FAAE in weight percent as presented in this study refers to the percentage of yields conversion recovered based on the theoretical yield.

3 Results and Discussion

3.1 Dialkyl Carbonates for Biodiesel Production

Dialkyl carbonates, known for their interesting chemistry, have long been studied by chemists as a versatile class of organic compounds to be used as lubricants and functional fluids [5]. However, utilization of dialkyl carbonates in non-catalytic supercritical process for biodiesel production has never been described earlier.

Table 1 presented critical temperatures and pressures of several dialkyl carbonates used in this study and the resulting reaction pressures at 300°C, when treated in batch-type system. It could be seen that the reaction pressure decreases with higher molecular weight of longer alkyl chains in the dialkyl carbonates. Near the critical point of dialkyl carbonates as sub/supercritical fluids, properties such as density, heat capacity and viscosity may vary greatly with only small changes in temperature and pressure [7]. Therefore, in order to ensure that the reaction condition is constant at the pre-determined optimized condition of 300°C/20 MPa throughout the experiment [7], the flow-type system was used, as both temperature and pressure, could be controlled.

Table 1 Critical temperatures and pressures of various dialkyl carbonates studied in this work

Dialkyl carbonates	Critical temperature (°C)	Critical pressure (MPa)	Pressure at 300°C (MPa)	Condition
Dimethyl carbonate (DMC)	275	4.6	10.3	Supercritical
Diethyl carbonate (DEC)	300	3.8	7.9	Supercritical
Dipropyl carbonate (DPC)	345	2.8	6.4	Subcritical
Dibutyl carbonate (DBC)	376	2.4	4.2	Subcritical

3.2 *Triglycerides Reactivity in Sub/Supercritical Dialkyl Carbonates*

Figure 1 shows the HPLC chromatograms as triglycerides were treated in dialkyl carbonates at 300°C/20 MPa/15 min. Those HPLC chromatograms showed peaks of the corresponding FAAE, which are different in retention times when treated in different dialkyl carbonates. As an example, methyl oleate from fatty acid methyl esters (FAME), obtained from supercritical DMC could be detected at 8.0 min retention time, while the others are detected later, at 9.0, 10.0 and 10.8 min for fatty acid ethyl esters (FAEE) from supercritical DEC, fatty acid propyl esters (FAPE) from DPC and fatty acid butyl esters (FABE) from DBC, respectively. In the meantime, the peaks of unreacted and intermediate compounds (monoglycerides and diglycerides) could be seen in chromatograms of FAPE and FABE, representing incomplete conversion from subcritical DPC and DBC.

In Fig. 2, dialkyl carbonates with shorter alkyl chains proved to yield more FAAE than the longer ones. Of all cases, the yield of FAAE increases over time. The highest of yield of all was FAME from DMC, achieving 97 wt.% after 15 min treatment, as previously reported [6]. Under a similar condition after 15 min, FAEE from DEC, FAPE from DPC and FABE from DBC produced 93, 70 and 65 wt.% yield, respectively. The high yield FAME from dimethyl carbonate over other dialkyl carbonates is expected due to the higher reactivity of its two methoxyl group [5]. At longer reaction time over 40 min, the yield of FAAE showed decreasing trends, due to thermal deterioration [7].

3.3 *Fatty Acids Reactivity in Sub/Supercritical Dialkyl Carbonates*

Several fatty acids, oleic, palmitic, linoleic, linolenic and stearic, were also reacted with various dialkyl carbonates in non-catalytic sub/supercritical condition similar as described beforehand. Those fatty acids showed significant conversion to its

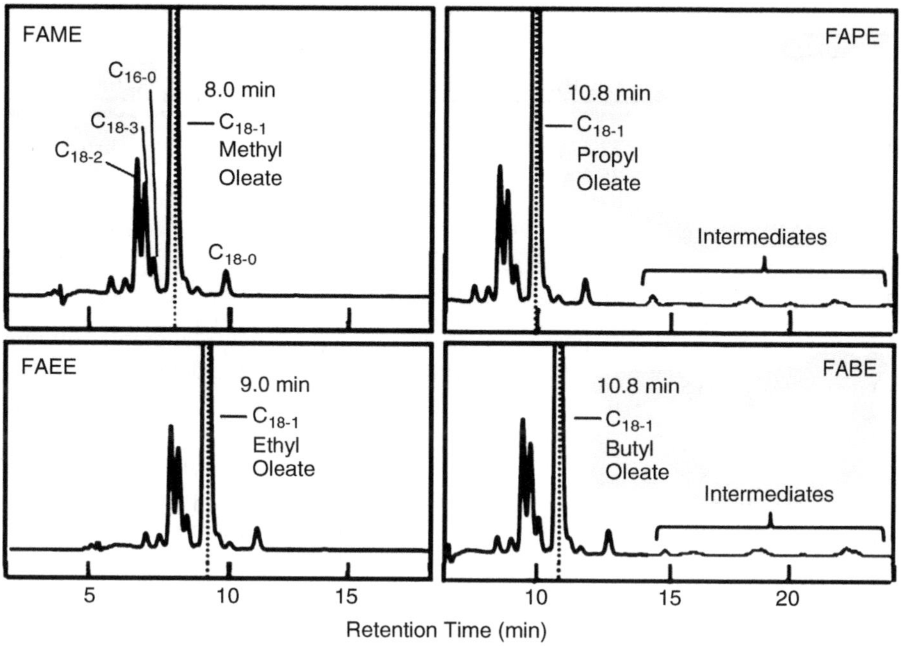

Fig. 1 HPLC chromatograms of the obtained FAAE after triglycerides treated in various sub/supercritical dialkyl carbonates at 300°C/20 MPa/15 min. First number represents carbon chain length while the latter shows number of double bonds

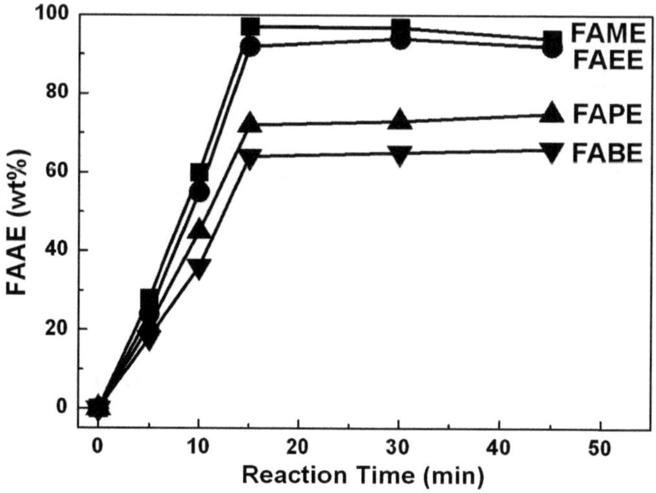

Fig. 2 FAAE yield from triglycerides treated in various sub/supercritical dialkyl carbonates at 300°C/20 MPa/15 min

Table 2 FAAE yield from various sub/supercritical dialkyl carbonates (300°C/20 MPa)

Reaction time (min)	Fatty acid					FAAE yield (mean wt.%)
	Oleic	Linoleic	Linolenic	Palmitic	Stearic	
DMC						
2	58.2	52.4	40.1	26.1	26.2	40.6
4	86.4	85.3	84.6	83.9	81.3	84.3
8	94.1	93.6	93.2	92.3	90.3	92.7
10	97.8	97.2	94.2	97.4	97.6	96.8
DEC						
2	37.6	36.8	37.3	36.5	35.3	36.7
4	76.2	75.3	73.2	71.5	71.3	73.5
8	87.2	85.3	86.1	83.4	79.5	84.3
10	96.5	96.5	93.5	97.3	97.2	96.2
DPC						
2	24.3	22.4	21.3	20.2	20.8	21.8
4	65.2	64.3	63.1	61.5	59.2	62.6
8	72.2	71.2	71.1	70.5	71.0	71.2
10	84.0	82.4	82.2	81.3	80.6	82.3
DBC						
2	19.2	18.5	17.6	12.2	13.4	16.2
4	44.8	42.3	43.4	41.6	38.3	42.1
8	60.3	58.4	58.9	58.4	57.1	58.6
10	66.5	67.1	66.8	66.9	66.2	66.7

corresponding alkyl esters in various dialkyl carbonates in shorter time than interesterification of triglycerides. Detailed results of the five fatty acids in supercritical dialkyl carbonates are presented in Table 2. In all cases, based on the HPLC analysis, peaks of fatty acids, appearing at 6.0–6.5 min retention time decreases, while their corresponding alkyl esters at 8.0–10.0 min retention time increase over time (data not shown).

The reaction time for complete conversion is the fastest for esterification in dimethyl carbonate after 10 min, increasing steadily with longer alkyl chains of dialkyl carbonates. Ten minute treatment was adequate to achieve high conversion of FAME formation for dimethyl carbonate. As expected and proven previously [4], unsaturated fatty acids (oleic, linoleic and linolenic acids) showed higher reactivity than those of saturated ones (palmitic and stearic acids).

3.4 *Potential By-Products from Sub/Supercritical Dialkyl Carbonates*

The highlight in the sub/supercritical dialkyl carbonates method is that no glycerol is produced, but instead, potential value-added by-products could be obtained. It should be noted that, however, FAAE from sub/supercritical dialkyl carbonates

Table 3 Expected products and by-products from sub/supercritical dialkyl carbonates

	R_A (alkyl residues in dialkyl carbonates)			
R_AOCOOR_A	Methyl (CH_3)	Ethyl (C_2H_5)	Propyl (C_3H_7)	Butyl (C_4H_9)
Dialkyl carbonates	Dimethyl carbonate (DMC)	Diethyl carbonate (DEC)	Dipropyl carbonate (DPC)	Dibutyl carbonate (DBC)
FAAE products	FAME	FAEE	FAPE	FABE
By-products	Glycerol carbonate etc.			

do not yet comply with international standards (>96.5 wt.%) for use as biodiesel except for the one from dimethyl carbonate. Further improvement or modification is still needed for the existing method. Based on those results from sub/supercritical dialkyl carbonates and published reaction equations [6, 7], potential products and value-added by-products were depicted in Table 3.

4 Conclusions

From the experimental evaluations made in this study, the reactivity of triglycerides and fatty acid in various sub/supercritical dialkyl carbonates was investigated. It obviously demonstrated that the esterification of fatty acids proceeds at higher reaction rate than the interesterification of triglycerides. Dialkyl carbonates with shorter alkyl chains also showed higher reactivity than the longer ones (dimethyl carbonate > diethyl carbonate > dipropyl carbonate > dibutyl carbonate).

In addition, the sub/supercritical alkyl carbonates method introduced in this study is a promising way to pioneer many other potential routes in producing biomass-derived chemicals to improve the current biodiesel production scenario.

Acknowledgement This study is partly funded by Japan Science and Technology Agency (JST) and Global-COE Program "Energy Science in the Age of Global Warming", Kyoto University, as well as Grant-in-Aid for Scientific Research "Challenging Exploratory Research" (April 2011 to March 2013), supported by Ministry of Education, Culture, Sports, Science and Technology-Japan, for all of which the authors highly acknowledge.

References

1. Mittelbach M, Remschmidt C (2004) Biodiesel: the comprehensive handbook. Boersedruck Ges.m.b.H, Vienna
2. Thompson JC, He BB (2006) Characterization of crude glycerol from biodiesel production from multiple feedstocks. Appl Eng Agric 22:261–265
3. Saka S, Isayama Y (2009) A new process for catalyst-free production of biodiesel by supercritical methyl acetate. Fuel 88:1307–1313

4. Saka S, Isayama Y, Ilham Z, Xin J (2010) New process for catalyst-free biodiesel production using subcritical acetic acid and supercritical methanol. Fuel 89:1442–1446
5. Parrish JP, Salvatore RN, Jung KW (2000) Perspectives on alkyl carbonates in organic synthesis. Tetrahedron 56:8207–8237
6. Ilham Z, Saka S (2011) Production of biodiesel with glycerol carbonate by non-catalytic supercritical dimethyl carbonate. Lipid Technol 23:10–13
7. Ilham Z, Saka S (2012) Optimization of supercritical dimethyl carbonate method for biodiesel production. Fuel 97:670–677

Transformation of Crystalline Cellulose III_I to Cellulose I_β in Semi-Flow Hot-Compressed Water Treatment

Rosnah Abdullah and Shiro Saka

Abstract Crystalline cellulose I (a mixture of I_α and I_β) from cotton linter (*Gossypium hirstum L.*) was converted into cellulose III_I by ethylenediamine treatment. However, the crystalline cellulose III_I was found to be transformed into cellulose I_β after semi-flow hot-compressed water (HCW) treatment at a range of temperatures between 20–270°C under 4–10 MPa for 15 min. This transformation, from cellulose III_I to I_β, was studied by X-ray diffraction (XRD) method. The analysis by the XRD method revealed that the transformation occurred at temperature below 200°C. The XRD peaks showed that the 002 lattice diffraction angles for treated crystalline cellulose III_I were shifted to around 22.5° from about 20.8°. These results confirmed that the conversion process was occurred in the semi-flow HCW system.

Keywords Cellulose • Cotton linter • Crystalline structure • Hot-compressed water • Hydrolysis

1 Introduction

Cellulose in its native form is generally in crystalline cellulose I structure. There exists two crystalline polymorphs in cellulose I; the well-oriented cellulose I_α and the highly crystalline cellulose I_β [1, 2]. The structures of celluloses I_α and I_β are assigned to the one-chain triclinic and two-chain monoclinic crystalline systems [3, 4], respectively, and the ratio of celluloses I_α and I_β in cellulose I varies according to the cellulosic material [5]. The I_β phase is considered to be more stable than the I_α phase because I_α phase is transformed into I_β after hydrothermal treatment.

R. Abdullah • S. Saka (✉)
Graduate School of Energy Science, Kyoto University, Yoshida-honmachi, Sakyo-ku, Kyoto 606-8501, Japan
e-mail: saka@energy.kyoto-u.ac.jp

T. Yao (ed.), *Zero-Carbon Energy Kyoto 2012*, Green Energy and Technology, DOI 10.1007/978-4-431-54264-3_11, © Springer Japan 2013

Another polymorph, cellulose III_I can be prepared by the treatment of cellulose I in liquid ammonia or various other amines followed by the removal of these reagents [6]. Cellulose III_I can be reversed and converted into cellulose I_β by treatment in warm water, whereas, cellulose III_I obtained under high pressure was only stable to boiling water for few hours [7–9]. Such conversion was also found to take place even without water by physical means [10].

Previously, cellulose III_I was observed to be converted into cellulose I after HCW treatment [11]. In this study, the reverse transformation was further examined under different pressure conditions in semi-flow HCW treatment in order to clarify the behavior of the cellulose III_I and to verify the stability of cellulose III_I in HCW treatment conditions.

2 Experimental

2.1 *Material*

The material used in this study was crystalline cellulose from cotton linter (*Gossypium hirstum L.*). It contains >90% of cellulose in the form of crystalline cellulose I [12]. This species was found to have about 80% mass fraction of I_β in its cellulose content with the rest being 20% of I_α [5, 13]. This linter was converted into cellulose III_I by ethylenediamine treatment [14]. After drying process, cellulose III_I samples were stored in a desiccator over silicon pellets until further usage.

2.2 *Hot-Compressed Water (HCW) Treatment and Analyses*

The semi-flow HCW system and its operational procedures as carried out formerly [15] were used in this study. Here, the cellulose III_I sample was treated at temperatures between 20–270°C under 4, 6 and 10 MPa for 15 min. The insoluble residues left in the reaction cell after treatment were collected, oven-dried and analyzed independently.

X-ray diffraction diagram: X-ray diffraction diagrams of the untreated samples and residues after HCW treatment were recorded using an imaging plate (IP) of 80 × 100 mm, mounted on the X-ray box of Rigaku RINT 2,200 V. Ni-filtered Cu–K_α radiation ($\lambda = 0.15418$ nm) generated at 40 kV and 30 mA was used. The samples were irradiated using an incident X-ray beam that was collimated in the box. The IP was then converted to X-ray diffraction diagrams via high sensitivity IP reader RAXIA-Di.

X-ray diffractometry: equatorial X-ray diffraction profiles were also obtained and the degree of crystallinity (%) was estimated accordingly [16].

FT-IR spctroscopy: the spectra of the sample pellets in KBr were recorded using a Shimadzu IR-8000 spectrophotometer. All spectra were recorded with an accumulation of 64 scans, revolution of 4 cm^{-1}, in a range from 4,000 to 400 cm^{-1}.

3 Results and Discussion

3.1 X-ray Diffraction of Cellulose III_I and I_β

Figure 1 showed the X-ray diffraction patterns of untreated and treated cellulose III_I. The pattern for cellulose III_I (left) has a strong reflections indexed as 011 and 002 located at the equator at $d = 0.75$ and 0.43 nm [17], respectively. Figure 1 (right) illustrated the X-ray diffraction diagram of the converted cellulose III_I to cellulose I_β at 270°C/10 MPa/15 min. The reflections indexed as 011, 011 and 002 occurred on the equator at $d = 0.60$, 0.56 and 0.40 nm, respectively.

Furthermore, all reflections in the diagram were indexed according to the I_β monoclinic unit cell as there was no characteristic I_α reflections observed [3]. This indicated that cellulose III_I fully transformed into cellulose I_β by the HCW treatment at 270°C/10 MPa/15 min. The X-ray diffractions for the transformation of cellulose III_I to I_β at other treatment conditions were not shown here, however found to be similar.

3.2 X-ray Diffraction (XRD) Profile of Cellulose III_I to I_β

X-ray diffraction profiles were plotted over various temperatures in order to observe the full transformation of cellulose III_I to cellulose I_β. For 10 MPa at 15 min

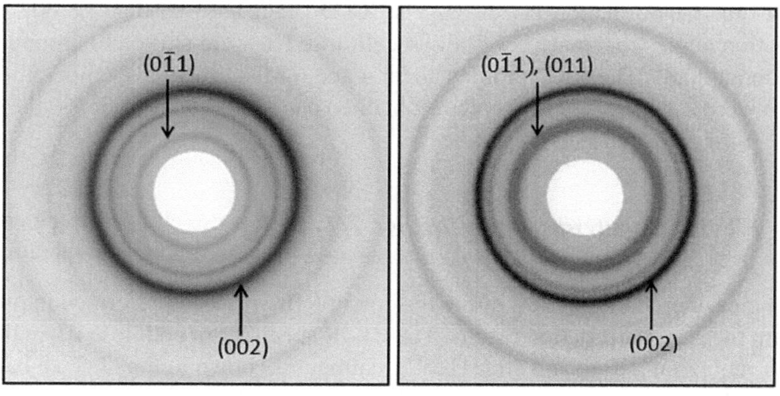

Fig. 1 X-ray diffraction diagrams of untreated crystalline cellulose III_I (*left*) and the treated cellulose III_I to be cellulose I_β by HCW at 270°C/10 MPa/15 min (*right*)

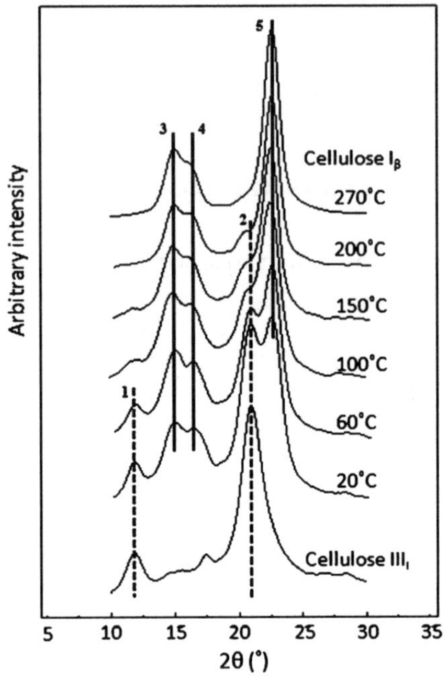

Fig. 2 XRD profiles for crystalline cellulose III_I, transforming to I_β as treated by HCW at various temperatures under 10 MPa/15 min

treatment time, the full transformation to cellulose I_β occurred at 270°C as shown in Fig. 2. Two equatorial reflections of cellulose III_I, labelled as 1 and 2 indexed at 011 and 002, respectively, were observed for the untreated cellulose III_I. As soon as cellulose III_I treated at very low temperature at 10 MPa for 15 min, a peak 2 shifted noticeably to the smaller angles because of thermal expansion, in accordance with the peak 1 being smaller. When the temperature increases to 100°C, the peak 1 slowly disappeared and peak 2 became smaller at 150°C and diminished. Whereas peaks 3–5 that correspond to cellulose I_β appeared markedly. Consequently, at 200°C, cellulose III_I was almost completely converted into cellulose I_β phase, with a full transformation at 270°C. The XRD peaks showed that the 002 lattice diffraction angles for treated crystalline cellulose III_I were shifted to around 22.5° from about 20.8°. The XRD profiles under 4 and 6 MPa at various temperatures for 15 min were not presented here, yet the obtained patterns were more or less the same.

3.3 FT-IR Spectrum of Cellulose III_I to I_β

FT-IR measurement of pellets of the residue in KBr was also performed in order to confirm the X-ray diffraction results. The FT-IR spectrum of cellulose III_I in the OH stretching region before and after HCW treatment is shown in Fig. 3. Two bands at 3,480 and 3,300 cm^{-1} were found in the spectrum of cellulose III_I, whereas the HCW-treated spectrum was found to be similar as that of *halocynthia* and ramie,

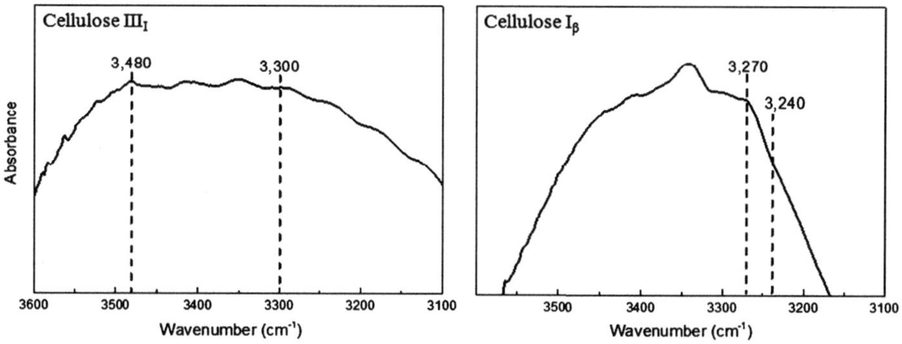

Fig. 3 The FT-IR spectra in the OH region of celluloses III_I and I_β at 270°C/10 MPa/15 min

Table 1 The change in degree of crystallinity of cellulose III_I transformed to cellulose I_β under different treatment conditions

	Crystallinity (%)								
		HCW treatment at various temperatures (°C)							
Pressure (MPa)	Untreated	60	80	100	150	180	200	230	270
4	84	89	89	90	90	91	–	–	–
6	84	88	89	88	91	91	91	92	–
10	84	89	88	88	90	90	92	91	92

which are I_β type cellulose [2, 10]. The bands at 3,240 and 3,270 cm^{-1} were assigned to I_α and I_β, respectively [18]. In the spectrum of the HCW treated cellulose, only the band at 3,270 cm^{-1} was clearly observed, but no bands were present at 3,240 cm^{-1}. Thus, the treated sample was cellulose I_β. A similar behavior was found for cellulose III_I treated by semi-flow HCW under 4 and 6 MPa for 15 min. The full transformation took place at 180°C/4 MPa/15 min and 230°C/6 MPa/15 min.

3.4 *Crystallinity Change Under Various HCW Treatment Conditions*

Table 1 showed the changes in degree of crystallinity of cellulose III_I transformed to cellulose I_β at various treatment conditions. It can be seen that there was a slight increase in crystallinity with the increase in temperature. This was perhaps due to the removal of paracrystalline parts in the cellulose and also annealing effect [19]. For 4 MPa, the highest temperature for HCW treatment to completely convert cellulose III_I to cellulose I_β was 180°C. While for 6 and 10 MPa, the highest temperatures were 230°C and 270°C, respectively.

4 Concluding Remarks

The result in literature [11] showed that cellulose III_I decomposed more extensively than cellulose I, showing that the conversion from cellulose I to cellulose III_I has some effect in HCW treatment. Therefore, the conversion from cellulose I to cellulose III_I for cellulose exploitation is required. On the other hand, it was also found that the cellulose III_I was more stable in HCW condition as compared with previous studies [7–9]. The semi-flow HCW treatment conditions at a range of temperatures between 20–270°C under 4–10 MPa for 15 min resulted a slower transformation of cellulose III_I to cellulose I_β. This result would be useful in analyzing the cellulose III_I behavior when it is treated in semi-flow HCW system, and as additional information to the studies mentioned earlier in literature. Nevertheless, further investigation would be needed in the future for more in details under various HCW treatment conditions for its transformation behaviors.

Acknowledgement The authors are grateful for the financial support provided under the GCOE program, Kyoto University.

References

1. Atalla RH, VanderHart DL (1984) Native cellulose: a composite of two distinct crystalline forms. Science 223:283–284
2. Zuganmaier P (2008) Crystalline cellulose and cellulose derivatives: characterization and structure. Springer, Heidelberg, pp 101–143
3. Sugiyama J, Okano T, Yamamoto H, Horii F, Odani H (1990) Experimental evidence for a triclinic system in native cellulose. In: Abstracts of papers of the American Chemical Society 199, Cell, part 1, Boston, 22 April, p 31
4. Sugiyama J, Vuong R, Chanzy H (1991) Electron diffraction study on the two crystalline phases occurring in native cellulose from an algal cell wall. Macromolecule 24:4168–4175
5. Horii F (2000) Structure of cellulose: recent developments in its characterization. In: Hon DNS, Shiraishi N (eds) Wood and cellulosic chemistry, 2nd edn. Dekker, New York, pp 51–81
6. Isogai A (1994) Allomorphs of cellulose and other polysaccharides. In: Gilbert RD (ed) Cellulosic polymers. Hanser, Munich, pp 1–24
7. Sueoka A, Hayashi J, Watanabe S (1973) The difference between natural cellulose I and regenerated cellulose I (I'). Nippon Kagaku Kaishi 3:594 (in Japanese)
8. Roche E, Chanzy H (1981) Electron microscopy study of the transformation of cellulose I into cellulose III_I in Valonia. Int J Biol Macromol 3:201–206
9. Yatsu LY, Calamari TA Jr, Benerito RR (1986) Conversion of cellulose I to stable cellulose III. Text Res J 56:419–424
10. Wada M (2001) In situ observation of the crystalline transformation from cellulose III_I to I. Macromolecule 34:3271–3275
11. Abdullah R, Saka S (2012) Hydrolysis behavior of various crystalline celluloses from cotton linter as treated by one-step semi-flow hot-compressed water. In: Yao T (ed) Zero-carbon energy Kyoto 2011. Springer, Tokyo, pp 127–132
12. Ishikawa Y, Saka S (2001) Chemical conversion of cellulose as treated in supercritical methanol. Cellulose 8:189–195

13. Wada M, Sugiyama J, Okano T (1995) Two crystalline phase (I_α/I_β) system of native celluloses in relation to plant phylogenesis. Mokuzai Gakkaishi 41:186–192
14. Isogai A, Usuda M, Kato T, Uryu T, Atalla RH (1989) Solid-state CP/MAS ^{13}C NMR study of cellulose polymorphs. Macromolecule 22:3168–3172
15. Lu X, Yamauchi K, Phaiboonsilpa N, Saka S (2009) Two-step hydrolysis of Japanese beech as treated by semi-flow hot-compressed water. J Wood Sci 55:367–375
16. Segal L, Creely JJ, Martin AE Jr, Conrad C (1959) An empirical method for estimating the degree of crystallinity of native cellulose using the X-ray diffractometer. Text Res J 29:786–793
17. Wada M, Heux L, Isogai A, Nishiyama Y, Chanzy H, Sugiyama J (2001) Improved structural data of cellulose III_I prepared in supercritical ammonia. Macromolecule 34:1237–1243
18. Sugiyama J, Persson J, Chanzy H (1991) Combined infrared and electron diffraction study of the polymorphism of native celluloses. Macromolecule 24:2461–2466
19. Yamamoto H, Horii F (1993) CP/MAS 13CNMR analysis of the crystal transformation induced for Valonia cellulose by annealing at high temperature. Macromolecule 26:1313–1317

Physico-Chemical Properties of Biodiesel from Various Feedstocks

Zan Win Moh Moh Phoo, Zul Ilham, Fadjar Goembira, Luis Razon, and Shiro Saka

Abstract In this study, plant oils were extracted from a wide variety of seeds to evaluate their potential as possible feedstocks for biodiesel production. The extracted oils were investigated by determining acid value, free fatty acids, tocopherol concentration, iodine value, density, water content and kinematic viscosity, etc. Afterwards, biodiesel was produced through esterification in the presence of methanol by using acid catalyst followed by transesterification with alkali catalyst, particularly when the FFA content of the oil was high. The biodiesel from these feedstocks were then analyzed for their physico-chemical properties such as acid value, tocopherol, iodine value, density, water content, kinematic viscosity, pour point, cloud point, flash point, cold filter plugging point, carbon residue content, oxidation stability, methanol content, and total glycerol, etc. Correlations between biodiesel properties were subsequently evaluated and the obtained results were discussed for improved biodiesel production.

Keywords Alternative feedstocks • Biodiesel • Oil and biodiesel properties

Z.W.M.M. Phoo
Graduate School of Energy Science, Kyoto University, Kyoto 606-8501, Japan

College of Engineering, De La Salle University, Manila, Philippines

Z. Ilham • F. Goembira • S. Saka (✉)
Graduate School of Energy Science, Kyoto University, Kyoto 606-8501, Japan
e-mail: saka@energy.kyoto-u.ac.jp

L. Razon
College of Engineering, De La Salle University, Manila, Philippines

T. Yao (ed.), *Zero-Carbon Energy Kyoto 2012*, Green Energy and Technology,
DOI 10.1007/978-4-431-54264-3_12,

1 Introduction

Biodiesel can be produced by transesterification of triglycerides and alcohols in the presence of an acid or an alkali catalyst and used as an alternative for diesel fuel. Biodiesel has become more attractive recently because of its environmental benefits and the fact that it is made from renewable biological sources [1]. Although both edible and non-edible oils can be used for biodiesel production, more than 95% of the world biodiesel production comes from edible oils at present [2]. The use of edible oils in biodiesel production may compete with food supply in the long term [3]. In contrast, production from non-edible oils does not compete with food application. Therefore, the production of biodiesel from different oilseed crops has been widely explored over the last few years [4–7].

In this research work, both edible and non-edible plant oils such as *Anacardium occidentale* (cashew), *Senna siamea* (cassia), *Ceiba pentandra* (kapok), *Leucaena leucocephala* (lead), *Albizia saman* (monkey pod), *Bombax ceiba* (red silk cotton), *Abelmoschus esculentus* (okra), *Calotropis gigantea* (Indian milkweed), *Annona muricata* (soursop), *Pongamia pinnata* (Indian beech) and *Jatropha curcas* (physic nut) were studied to evaluate their potential as possible feedstocks for biodiesel production.

2 Experimental

The seeds used in this study were collected from Myanmar, Philippines, Malaysia, Indonesia and Thailand. The oils were extracted with hexane in a Soxhlet apparatus at 60°C for 8 h. Fatty acid composition of plant oils was analyzed using High Performance Liquid Chromatography (HPLC) (Shimadzu LC-10AT) equipped with Cadenza CD-C18 column and refractive index detector RID-10A. The column temperature was 40°C and methanol was used as a carrier solvent with a flow-rate of 1 mL/min. The free fatty acid content in oil was determined using AOCS Official Method AOCS Cd 3d-63. Tocopherol concentration was analyzed using HPLC equipped with Shodex Asahipak ODP-50 6D column and ultraviolent detector SPD-10A. Iodine value was measured according to WIJS method, while density was determined using I-type hydrometer. Karl Fischer moisture titrator MKC-520 (Kyoto Electronics MFG. Co. Ltd.) was used to check water content and Automatic Kinematic Viscosity Measuring System AKV-201 was used to test kinematic viscosity at 40°C.

Biodiesel was produced from these oils through acid-catalyzed esterification with methanol followed by alkali-catalyzed transesterification, particularly when the Free Fatty Acids (FFA) content of the oil was high. Both acid-catalyzed esterification and alkali-catalyzed transesterification were carried out in a round bottom flask mounted with condenser using oil bath stirrer for 1 h at 60–65°C.

Methanol and fatty acid methyl ester content were analyzed by Gel Permeation Chromatography (GPC) (Column: GF-310HQ, oven temperature: 40°C, flow-rate: 1 mL/min, mobile phase: acetone, detector: RID 10A). MP™ Saftest analyzer was used to determine total glycerol. Acid value was determined by AOCS Official Method AOCS Cd 3d-63. Tocopherol, iodine value, water content, density and kinematic viscosity were measured using the same method with oil properties determination. Oxidation stability was determined by Rancimat 743 (Methrom, Herisau, Switzerland), while carbon residue was measured by Micro Carbon Residue Tester ACR-M3. For the pour point and cloud point, the tests were made by a Mini Pour/Cloud Point Tester MPC-102. Automated Cold Filter Plugging Point Tester AFP-102 was used to determine cold filter plugging point, while Pensky-Martens Closed Cup Automated Flash Point Tester APM-7 was used to determine the flash point.

3 Results and Discussion

3.1 Physical and Chemical Properties of Plant Oils

Table 1 shows characteristic properties of the oils extracted from various plants for biodiesel production. Cashew has a good source of oil since its nut contains the highest oil content with 46.2 wt.% among these seeds. Indian milkweed, Indian beech and physic nut seeds comprise the higher oil content; 33.3, 39.2 and 45.0 wt.%, respectively. Kapok, okra, red silk cotton and soursop have intermediate levels of oil content, while cassia, lead and monkey pod have low oil content.

The FFA content in the oil should be less than 0.5 wt.% and all materials should be substantially anhydrous prior to alkali-catalyzed transesterification [8]. FFA contents in the oils used in this study were greater than 0.5 wt.% and the range of water content was 65–2,513 mg/kg (0.0065–0.25 wt.%). Therefore, two-step process; acid-catalyzed esterification and alkali-catalyzed transesterification were conducted. Fatty acid composition of plant oils is as shown in Fig. 1.

3.2 Physico-Chemical Properties of Biodiesel

Characteristic fuel properties of biodiesel prepared from various plant oils by alkali-catalyzed method is described in Table 2. One of the important criterions for a good-quality biodiesel is the completion of the transesterification reaction. Total glycerol measures the amount of unconverted or partially converted fats and by-product glycerol present in the fuel. Total glycerol of cashew, cassia, kapok, lead, okra, Indian milkweed, soursop and Indian beech was lower than the international biodiesel standard value 0.24 wt.% so it cannot cause negative effects on the

Table 1 Characteristic properties of the oils extracted from various plants for biodiesel production

Sample no	Name of plant	Oil content in seeds (wt%)	Free fatty acid content in oils (wt%)	Tocopherol concentration (ppm)	Iodine value (gI_2/100 g)	Density (Kg/m^3) (15°C)	Water content in oil (mg/kg)	Kinematic viscosity (mm^2/s) (40°C)
1	Cashew	46.2	9.1	112	93.6	0.92	1,587	36.4
2	Cassia	5.4	5.5	1,772	102.5	0.93	2,513	36.6
3	Kapok	23.1	21.5	855	81.9	0.92	1,801	28.7
4	Lead	4.2	3.0	1,875	112.4	0.93	1,673	37.0
5	Monkey pod	4.7	1.3	2,353	103.3	0.93	1,630	38.0
6	Okra	17.0	0.76	699	110.5	0.92	213	29.2
7	Red silk cotton	20.2	65.7	1,719	88.9	0.89 (45°C)	874	21.6
8	Indian milkweed	33.3	27.5	443	102.2	0.92	1,739	25.8
9	Soursop	20.5	3.5	1,368	109.7	0.93	65	38.1
10	Indian beech	39.2	2.2	385	113.0	0.93	700	38.5
11	Physic nut	45.0	13.6	1,653	103.6	0.92	313	48.0

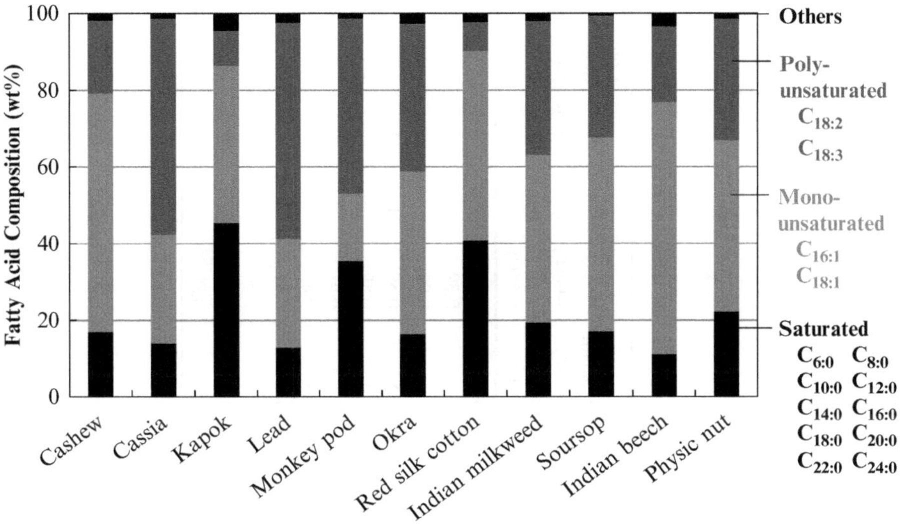

Fig. 1 Fatty acid composition of plant oils

diesel engine, such as clogging fuel filters, fouling fuel injectors, and forming sediment on the bottom of fuel storage tanks. However, total glycerol of monkey pod, red silk cotton and physic nut was above the standard value. Consequently, the biodiesel produced from those plant oils needs to be washed out carefully since washing step can remove total glycerol during the biodiesel production.

It was found that okra biodiesel obtained the highest conversion factor with ester content up to 99.9 wt.%. Similarly, those of cassia, kapok, lead, Indian milkweed and soursop biodiesel were 99.7, 99.8, 99.6, 99.6 and 99.7 wt.%, respectively. Indian beech and physic nut biodiesel also occupied amount of fatty acid methyl ester content up to 97.9 and 97.5 wt.%. But, its ester content of monkey pod biodiesel was the lowest value of 18.9 wt.% while that of red silk cotton biodiesel showed 58.3 wt.% due to its higher free fatty acids in oils. The lower acid value and carbon residue obtained in this study suggested that incomplete combustion would not occur in the biodiesel.

Oxidation stability is one of the most important fuel properties relating to in-use performance of biodiesel. Oxidation stability is influenced by the fatty acid profile of plant oils. The higher unsaturation normally leads to the poorer stability. Monkey pod biodiesel has the higher oxidation stability with 6.5 h as its plant oil composes of a large amount of saturated fatty acids. Correspondingly, soursop biodiesel has the lowest oxidation stability with 0.6 h since its plant oil contains 82.4 wt.% unsaturated fatty acids. The lower the iodine value, the lower the NOx emissions [9]. It was learned that the iodine value obtained in this study cannot give significant effects on NOx emissions as iodine values of all biodiesels comply with the international standard range ($120 gI_2/100$ g at maximum).

Table 2 Characteristic fuel properties of biodiesel prepared from various plant oils by alkali-catalyzed method

Sample no	Total glycerol (wt%)	Methanol (wt%)	Ester content (wt%)	Carbon residue (wt%)	Acid value (mg KOH/g)	Oxidation stability (h)	Iodine value (gI_2/ 100 g)	Tocopherol concentration (ppm)	Water content (mg/kg)	Kinematic viscosity mm^2/s (40°C)	Density (kg/m^3) (15°C)	Pour point (°C)	Cloud point (°C)	Cold filter plugging point (°C)	Flash point (°C)
1	0.21	n.d.	92.9	0.01	0.25	5.6	82.9	59	683	4.6	0.88	3	4	3	113
2	0.10	n.d.	99.7	0.03	0.27	3.9	65.4	1,331	145	4.0	0.89	7	7	4	165
3	0.13	n.d.	99.8	0.03	0.24	0.8	88.3	479	1,308	4.2	0.89	1	2	−4	167
4	0.11	n.d.	99.6	0.03	0.27	1.7	111.8	976	429	4.5	0.89	11	11	20	163
5	0.54	n.d.	18.9	0.32	0.23	6.5	94.6	1,613	597	17.0	0.89	16	16	−1	142
6	0.07	n.d.	99.9	0.00	0.22	1.6	89.8	141	59	4.1	0.87	6	6	6	126
7	0.58	n.d.	58.3	0.23	0.26	5.0	74.0	2,603	117	6.8	0.89	3	6	−2	116
8	0.05	n.d.	99.6	0.04	0.26	1.4	89.7	214	105	4.0	0.87	8	8	8	128
9	0.08	n.d.	99.7	0.08	0.27	0.6	84.2	50	76	4.8	0.88	−3	1	−1	118
10	0.22	0.18	97.9	0.23	0.43	4.5	87.8	375	340	4.6	0.89	−3	−1	−3	153
11	1.06	n.d.	97.5	0.53	1.05	6.7	82.0	1,233	754	5.3	0.88	−2	1	−2.5	173

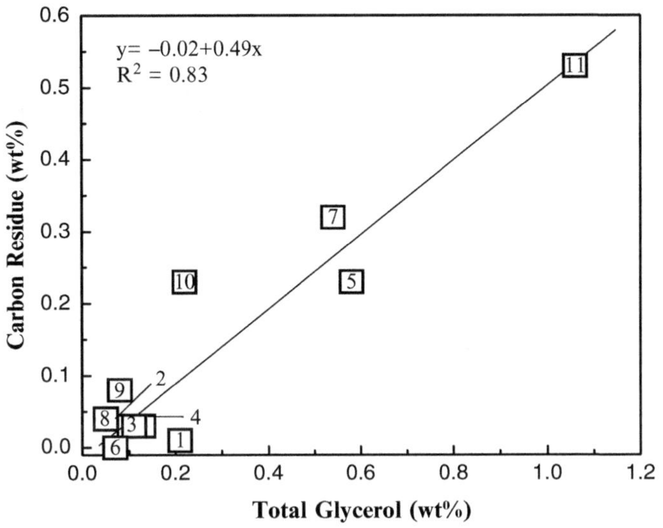

Fig. 2 Correlation between total glycerol and carbon residue

Viscosity is the major reason why the fats and oils are converted to biodiesel. The viscosity of biodiesel is approximately an order of magnitude lower than that of the patent oil or fat, leading to better atomization of the fuel in the combustion chamber of the engine. Fatty acid methyl ester of monkey pod and red silk cotton have higher viscosity with 17 and 6.8 mm^2/s respectively that make them not suitable to be used as biodiesel. In terms of clod flow properties, cashew, cassia, kapok, lead, monkey pod, okra, red silk cotton and Indian milkweed biodiesels are better to use in tropical countries. The flash point is influenced by the level of unreacted methanol remaining in the finished biodiesel. The biodiesel of cassia, kapok, lead, Indian beech and physic nut have higher flash point than that of cashew, monkey pod, okra, red silk cotton, Indian milkweed and soursop. This indicates that cassia, kapok, lead, Indian beech and physic nut biodiesels are safer in storage and handling.

3.3 Correlations Between Biodiesel Properties

The correlations between biodiesel properties were also studied. Figures 2 and 3 show correlation between total glycerol and carbon residue and relation between iodine value and oxidation stability.

The strength of the linear association between two variables is quantified by the correlation coefficient (R^2). Since its value was 0.83 in Fig. 2, it could be suggested that the total glycerol was highly correlated with carbon residue. Consequently, higher total glycerol leads to the higher value of carbon residue due to the

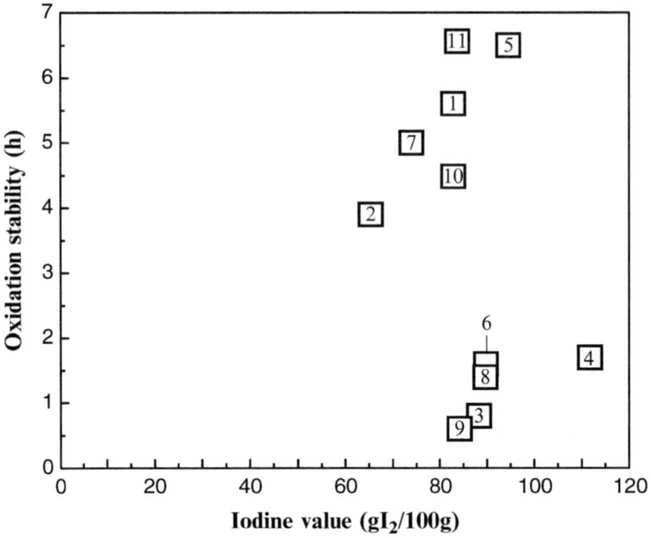

Fig. 3 Relation between iodine value and oxidation stability

incomplete combustion of the remaining glycerol in the biodiesel. In addition, oxidation stability is influenced by unsaturation. The autoxidation of unsaturated compounds proceeds at different rates depending upon the number and position of the double bonds. Two groups of data could be seen in Fig. 3. Oxidation stability of biodiesel was observed to be higher with increase in iodine value for both groups due to variation of unsaturated fatty acids measured by degree of unsaturation.

4 Conclusions

Based on the biodiesel properties, biodiesel produced from monkey pod and red silk cotton showed poor physico-chemical properties and low ester content. However, those of cashew, cassia, kapok, lead, okra, Indian milkweed, soursop, Indian beech and physic nut displayed high potential to be used as biodiesel. As a conclusion, biodiesel production process must be made watchfully, especially for monkey pod and red silk cotton oils.

Acknowledgement The authors are grateful for the financial support provided by AUN/SEED Net Project (JICA).

References

1. Singh SP, Singh D (2010) Biodiesel production through the use of different sources and characterization of oils and their esters as the substitute of diesel: a review. Renewable Sustain Energy Rev 14:200–216
2. Gui MM, Lee KT, Bhatia S (2008) Feasibility of edible oil vs non-edible oil vs waste edible oil as biodiesel feedstock. Energy 33:1646–53
3. Refaat AA (2009) Correlation between the chemical structure of biodiesel and its physical properties. Int J Environ Sci Technol 6:677–94
4. Moser BR (2009) Biodiesel production, properties and feedstocks. In Vitro Cell Dev Biol Plant 45:229–266
5. Chhetri AB, Tango MS, Budge SM, Watts KC, Islam MR (2008) Non-edible plant oils as new sources for biodiesel production. Int J Mol Sci 9:169–80
6. Nakpong P, Wootthikanokkhan S (2010) Roselle (*Hibiscus sabdariffa L*) oil as an alternative feedstock for biodiesel production in Thailand. Fuel 89:1806–1811
7. Rashid U, Anwar F, Moser BR, Knothe G (2008) *Moringa oleifera* oil: a possible source of biodiesel. Bioresour Technol 99:8175–8179
8. Demirbas A (2009) Progress and recent trends in biodiesel fuels. Energy Convers Manag 50:14–34
9. Graboski MS, McCormick RL, Alleman TL, Herring AM (2003) The effect of biodiesel composition on engine emissions from a DDC series 60 diesel engine. Natl Renewable Energy Lab; NREL/SR-510-31461, p 52

Woodfuel Utilization for Cooking in Indonesian Household: Assessment of Indoor PM_{10} Pollution and Wood Consumption at National Level

Haryono S. Huboyo, Susumu Tohno, and Puji Lestari

Abstract After national fuel switching was launched in mid-2007 across Indonesia, people in rural area seem to have dual fuels (wood and LPG) for daily cooking. Field measurements of indoor air pollution related to cooking activities were accomplished in 40 households at two sites (Lembang and Juwana) of Indonesian rural areas during December 2010 to January 2011. Size segregated PM_{10} indoor aerosols related to cooking emission were quantified with a Sioutas Cascade impactor. The average PM_{10} concentrations related to the stoves emissions were $1{,}209 \pm 918\ \mu g/m^3$ in Lembang and $1{,}375 \pm 884\ \mu g/m^3$ in Juwana. Averages of 3.3 ± 1.6 kg and 3.8 ± 1.9 kg wood were consumed daily for Lembang site and Juwana site respectively. The PM_1 contributed to 86% and 88% of PM_{10} mass concentration at Lembang and Juwana respectively. While the $PM_{2.5}$ shared 90% and 92% of PM_{10} mass concentration at Lembang and Juwana respectively. Additionally, combination of measured unit consumption of woodfuel with the previous data on woodfuel consumption allowed national household consumption for cooking by island in Indonesia. The total consumption was estimated to be 106 million m^3/year and Java island accounted for the largest share of 31%.

Keywords Cooking stove • Energy • Fine particles • PM_1 • $PM_{2.5}$

H.S. Huboyo (✉) • S. Tohno
Department of Socio Environmental Energy Science, Graduate School of Energy Science, Kyoto University, Yoshida Honmachi Sakyo-ku, Kyoto, Japan
e-mail: haryono.huboyo.67w@st.kyoto-u.ac.jp

P. Lestari
Faculty of Civil and Environmental Engineering, Institute of Technology Bandung, Bandung, Indonesia

T. Yao (ed.), *Zero-Carbon Energy Kyoto 2012*, Green Energy and Technology,
DOI 10.1007/978-4-431-54264-3_13,

1 Introduction

Low income households in developing countries still rely on old biomass fuels, i.e. wood, charcoal and other solid fuels. In 2005, the share of woodfuel consumption for Asia-Pacific region was 41% (equivalent to 787 million m^3/year) of the total world consumption [1]. The survey conducted by the Ministry of Health, Indonesia showed that at least 17.3% of urban households and 64.2% of rural households still use biomass fuels, i.e. wood and charcoal [2] for their cooking energy. Woodfuel was gathered from the forest, garden, plantation and agricultural waste. Historically during 2000–2010, the woodfuel users at national level increased by 0.7% where household consumption itself accounts for 84.5% of the increment [3]. In addition, there are indications of additional users of woodfuel who were originally kerosene users after kerosene subsidy phase out program [4]. However the percentage of this shift has not been known yet. Practically, there are open options to choose cleaner fuel among many candidates and woodfuel continues to be an attractive option for the households.

Unit consumption of fuel wood in rural developing countries (after converted to m^3/cap/year unit) is in the range of 0.58–1.39 m^3/cap/year [5–8]. This consumption may change with urbanization, household income and wood accessibility. In Indonesia case, the studies on household wood consumption were scattered. However until now, there are no full scale studies to reveal unit consumption of woodfuel across the country. Adopting dual fuel (wood and LPG) is currently common for part of urban as well as rural households to secure their energy needs. This is partly due to the unstable subsidized LPG stock in terms of price or availability in the retail market.

Use of woodfuel for cooking poses health consequences for the cooks. Fullerton [9] listed respiratory and non-respiratory diseases related to biomass usage for both children and adults. It is associated with acute lower respiratory tract infection, low birth weight, nutritional deficiency (for children), chronic obstructive lung disease, tuberculosis, lung cancer, cardiovascular disease and cataracts (for adults). Since biomass is widely used in most of rural areas, it is expected that high exposure of housewives and children to biomass emission causes long-term health effects. For instance Tana et al. [10] revealed that cataract was prevalent for Indonesian housewives using biomass fuel rather than gas and kerosene fuel. The main cause of potential harm of biomass emission is fine particles which are generally within inhalable size range. PM_{10} and $PM_{2.5}$ are the two parameters that have been typically adopted for aerosol characterization. Nowadays, ultrafine particles (<100 nm) are more important with respect to acute effects of asthma [11]. Conventional woodstove combustion generally emits high proportion of fine particles mass (representing by high $PM_{2.5}$ mass). Moreover, the size distribution of emitted particles can be altered by modifying stove design. For example, using chimney will reduce fine fraction of the emission [12].

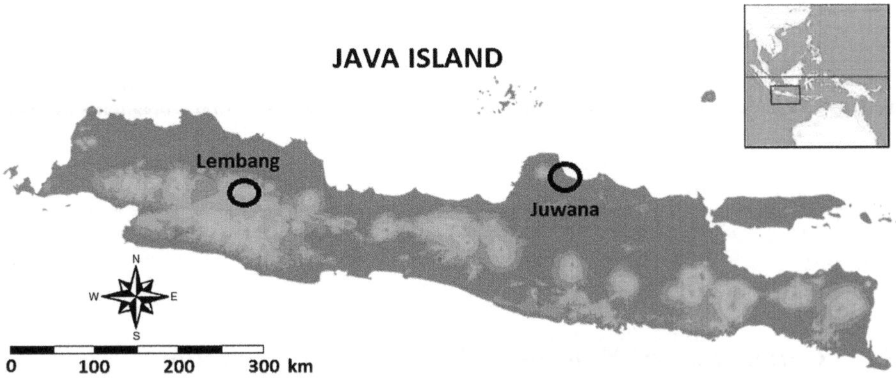

Fig. 1 The sampling sites for measurements

This study is aimed at quantifying three size-segregated mass size distributions (PM_1, $PM_{2.5}$ and PM_{10}) of aerosols related to cooking emission and estimating the demand of woodfuel for cooking energy of Indonesian households.

2 Methodology

2.1 Measurement of Particulate Matter Concentration in Household

The study was undertaken simultaneously at two different locations representing rural high-altitude and low-altitude regions in Indonesia. The highland rural area (1,280 m ASL) was in Lembang (Sunten Jaya village), West Java Province, and the lowland area (2 m ASL) was in Juwana (Bakaran Wetan village), Central Java Province (see Fig. 1). In each site, we measured 20 households which use woodfuel as the main fuel, however only 15 samples and 14 samples were valid for Lembang and Juwana respectively. In these areas, the households use usually woodfuel side by side with LPG fuel for their daily cooking. Using this scheme, it is expected the health effects on the cooks will still be pronounced. Therefore it is important to measure particulate matter concentrations related to biomass burning emission in these locations.

The mass size distribution of $PM_{2.5}$ was measured by a SKC Sioutas Cascade Impactor (SKC Inc) combined with a Leland Legacy® sample pump at constant flowrate of 9 L/m. This pump was calibrated using a Defender (model 510H Dry Cal Inc). The sampler was positioned 30–40 cm from stove perimeter to closely investigate indoor particles related emission. Collected particles on the filters were analyzed gravimetrically using a micro-balance with 1 μg precision (MX-5 model, Mettler Toledo Inc).

Table 1 Unit consumption (UC) of woodfuel used in this study (m^3/cap/year)

Island/(cluster)	UC	Sources	Island/(cluster)	UC	Sources
Sumatra	1.06	National average	Sulawesi	1.61	[13]
Java	0.88	Average of this study and [14–16]	Maluku	1.06	National average
Bali-N. Teng	1.06	National average	Papua	10.9	[17]
Kalimantan	1.01	[15]			

Note: Papua was excluded from national average calculation

2.2 Calculation of Household Woodfuel Consumption

The woodfuel consumption was evaluated on a daily basis. This data was then modified to obtain annual unit consumption per capita. We estimated the annual consumption of woodfuels in both urban and rural areas. Generally the wood fuel users in both areas are not necessarily categorized as low income people. According to government definition (in 2011), the low income people in urban area had an income less than 27.2 USD/month meanwhile in rural area their income were less than 22.95 USD/month (using current exchange rate 1 USD = Rp. 9,300). About 15% people in urban areas are categorized as low income households while around 9% people are wood users. In rural areas, about 69% people are wood users while people categorized as low income households are only 17%. The formula to calculate total woodfuel consumption is as follows:

$$\sum_{i}\left\{\mathrm{Cr_i}\ \mathrm{x}\left(\sum_{m}\mathrm{Fr_{i,m}Pr_{i,m}}\right)\right\}+\sum_{i}\left\{\mathrm{Cu_i}\ \mathrm{x}\left(\sum_{m}\mathrm{Fu_{i,m}Pu_{i,m}}\right)\right\} \quad (1)$$

where Cr: unit (per capita) consumption of woodfuel in rural areas (m^3/person/year); Cu: unit (per capita) consumption of woodfuel in urban areas (m^3/person/year); Fr: fraction of rural people using wood fuel; Fu: fraction of urban people using wood fuel; Pr: rural population; Pu: urban population; i: island or island cluster; m: province.

The unit consumption used here and the related sources are shown in Table 1. The assumptions used for estimation are: concerned biomass is addressed to woodfuel only, unit consumption of woodfuel was same in each island rather than each province, and wood density was 600 kg/m^3.

For ease, we will show the results on island basis. In urban area, biomass consumption per capita was adopted from ESMAP surveys [18] in urban developing countries including Indonesia (middle income: 0.11 m^3/cap/year, middle-low income: 0.14 m^3/cap/year). To determine the income group for each province we took into account minimum wage per month by province and gross regional domestic product by province [19].

3 Results and Discussion

3.1 *Assessment of Indoor PM Pollution Related to Woodfuel Burning for Cooking*

Cooking stoves used in the two sites have a comparable design of U-shape. The difference is that in Lembang site, the stoves are rectangular U-shape and made of cone block, while in Juwana the stoves were rounded U-shape and made of concrete. These stove types have differences on the hole at which the pot being placed and the shape of outer stove (Fig. 2). During measurements, the average temperature and relative humidity in the kitchen were 23 ± 2°C and 77 ± 6% (in Lembang) and 27 ± 1°C and 86 ± 3% (in Juwana).

The average PM_{10} concentrations related to the stoves emissions were 1,209 ± 918 μg/m^3 in Lembang and 1,375 ± 884 μg/m^3 in Juwana (Table 2). The average mass fractions of PM_1 and $PM_{2.5}$ to PM_{10} were 86 ± 5% and 90 ± 3% for Lembang. While in Juwana, the fractions were 88 ± 4% and 92 ± 3%. Our mass fraction results were comparable to other study (using the same device) related to open burning emission [12]. Daily wood fuel consumptions were 3.3 ± 1.6 kg/day in Lembang and 3.8 ± 1.9 kg/day in Juwana. In terms of unit consumption, these values correspond to 0.57 and 0.68 m^3/cap/year, respectively. In most cases, our PM_{10} exposure estimations for cooks showed lower values than those in Bolivia (i.e. 940–1,260 μg m^{-3} h^{-1}) by Albalak et al. [20].

3.2 *Assessment of Woodfuel Consumption in Indonesia*

Calculation results of the household woodfuel consumption (see Table 3) revealed that Java island stood as the highest percentage of woodfuel consumption due to high percentage of woodfuel user (15% in urban, 66% in rural) as well as the largest

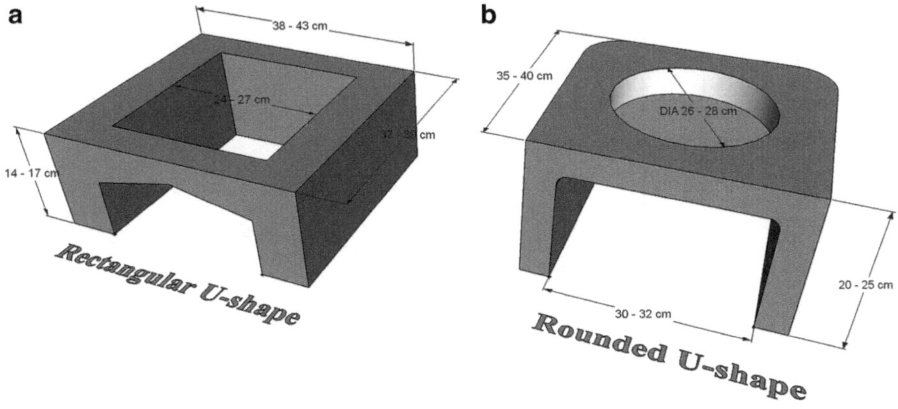

Fig. 2 Typical stoves used in two sampling sites

Table 2 PM_{10} concentrations, biomass consumed and exposure estimate

Sample (Lembang)	PM_{10} concentration (μg/m^3)	Biomass consumed (kg/day)	Exposure estimate μg/(m^{3}h)	Sample (Juwana)	PM_{10} concentration (μg/m^3)	Biomass consumed (kg/day)	Exposure estimate μg/(m^{3}h)
L1	2,292	3.3	724	J1	1,383	5.3	494
L2	1,831	3.6	631	J2	366	2.2	179
L3	1,280	2.4	422	J3	1,170	3.5	816
L4	3,253	1.2	1,574	J4	946	4.9	293
L5	1,520	5.4	512	J5	1,326	7.2	324
L6	892	1.6	357	J6	1,437	0.8	2,156
L7	1,552	3.3	711	J7	1,012		542
L8	2,349	3.3	1,146	J8	1,187	6.6	1,003
L9	619		167	J9	902	2.4	712
L10	432	2.0	204	J10	526	2.4	272
L11	199	5.5	55	J11	3,690	2.4	1,800
L12	271	1.5	85	J12	2,508	4.5	1,031
L13	808	2.4	288	J13	640	4.4	315
L14	726	6.7	276	J14	2,158	2.3	1,823
L15	124	3.7	41				

Note: the exposure was estimated at maximum case (based on cooking period, not activity time of the cooks)

Table 3 Calculated national biomass consumption (1,000 m^3/year) and deforestation rate/DR (% per year)

Island (cluster)	Urban	Rural	Classified FA (ha/capita)	DR	Unclassified FA (ha/capita)	DR
Sumatra	245	20,060	0.52	0.53	0.92	0.41
Java	1,504	31,016	0.03	0.09	0.22	0.08
Bali-Nusa Tenggara	190	7,198	0.19	0.03	0.56	0.02
Kalimantan	58	5,330	1.92	0.12	2.12	0.94
Sulawesi	103	13,392	0.70	0.04	0.73	0.16
Maluku	19	1,545	1.33	0.02	0.40	0.09
Papua	12	25,070	7.74	0.09	0.08	1.70
Indonesia	*2,133*	*103,611*	*0.55*	*0.18*	*0.57*	*0.41*

Note: Classified Forest Area (FA) comprises of conserve forest and protected forest only (production forests were not included). The capita of interest is addressed to wood user only. The deforestation rate (DR) was recalculated based on 2006–2009 data [21]

population. The second position was Papua island owing to extremely high biomass unit consumption. Papua island is the least developed island in Indonesia leading to importance of forest stock as the main cooking energy resources. Table 3 indicates that the total forest area per person of wood user varies and Java island shows the lowest. Therefore it is expected that woodfuel users (particularly for urban low income) in Java island will suffer the most resource scarcity in the future as a result of deforestation across the country.

Our estimated total consumption (106 million m^3/year) is much lower than national household consumption calculated by the Ministry of Energy, i.e., 175 million m^3/year [3] mostly due to the difference of unit consumptions used. Their unit consumptions rely on subjective quantification of the respondents for large-scale questionnaire survey and the accuracy is low. By contrast we rely on objective quantifications based on the limited field measurements. Moreover, we measured only for woodfuel that might not be specifically addressed in the Ministry of Energy data sources. To improve the accuracy, the unit consumption on mass basis of used woodfuel shall be considered by the Ministry of Energy in calculating biomass consumption at nationwide. Also the unit consumption of biomass should be updated after kerosene to LPG conversion program was launched in 2007.

4 Conclusion

The average mass fractions of PM_1 and $PM_{2.5}$ to PM_{10} were $86 \pm 5\%$ and $90 \pm 3\%$ for Lembang. While in Juwana, the fractions were $88 \pm 4\%$ and $92 \pm 3\%$. The average exposures to the cooks at Juwana were higher almost two folds (839 ± 651 $\mu g\ m^{-3}\ h^{-1}$) than in Lembang site (479 ± 429 $\mu g\ m^{-3}\ h^{-1}$). Java population still consumes wood fuel at the greatest nationally, however resource scarcity in the future will threat to the wood users due to deforestation.

Acknowledgment This work was financially supported by Global Centre of Excellence (GCOE) Program of Graduate School of Energy Science, Kyoto University for equipments and articles, also Environmental Management Leader Program of Kyoto University for travel expenses.

References

1. Gumartini T (2009) Biomass energy in the Asia-Pacific Region: current status, trends and future setting Asia-Pacific forestry sector outlook study. Working Paper No. APFSOS II/WP/ 2009/26, FAO Regional Office for Asia and the Pacific, Bangkok
2. National Institute of Health Research & Development (2010) Basic health research, Ministry of Health Republic of Indonesia
3. Center for Data and Information on Energy and Mineral Resources (2011) Handbook of energy & economic statistics of Indonesia ministry of energy and mineral resources. http://prokum.esdm.go.id/Publikasi/Handbook%20of%20Energy%20&%20Economic%20Statistics%20of%20Indonesia%20/Handbook%20of%20Energy%20&%20Economic%20Statistics%20ind%202011.pdf. Accessed 9 May 2012
4. Dwiprabowo H (2010) Study of policy on firewood as source of energy in rural areas in Java. Jurnal Analisis Kebijakan Kehutanan 7:1–11
5. Nansaior A, Patanothai A, Rambo AT, Simaraks S (2011) Climbing the energy ladder or diversifying energy sources? The continuing importance of household use of biomass energy in urbanizing communities in Northeast Thailand. Biomass Bioenergy 35:4180–4188
6. Wijesinghe LCAS (1984) A sample study of biomass fuel consumption in Sri Lanka households. Biomass 5:261–282
7. Rojas DT, Lehmann J, Hobbs P, Joseph S, Neufeldt H (2011) Biomass availability, energy consumption and biochar production in rural households of Western Kenya. Biomass Bioenergy 35:3537–3546
8. Jashimuddin M, Masum KM, Salam MA (2006) Preference and consumption pattern of biomass fuel in some disregarded villages of Bangladesh. Biomass Bioenergy 30:446–451
9. Fullerton DG, Bruce N, Gordon SB (2008) Indoor air pollution from biomass fuel smoke is a major health concern in the developing world. Trans R Soc Trop Med Hyg 102:843–851
10. Tana L, Delima D, Kristanto AY (2009) The association of fuel Use and cataract among housewives in Indonesia: national basic health research 2007. Majalah Kedokteran Indonesia 59:363–369
11. Air Quality Sciences (2011) Ultrafine particle: why all the concern about something so small? Air Quality Sciences, Inc. http://www.abatement.com/pdf/ultrafine-particles-why-the-concern.pdf. Accessed 15 Jun 2012
12. Armendariz AC, Edwards RD, Johnson M, Rosas IA, Espinosa F, Masera OR (2010) Indoor particle size distributions in homes with open fires and improved patsari cook stoves. Atmos Environ 44:2881–2886
13. NFP-FAO- Pattimura University (2012) Studi penguatan hak masyarakat dalam pengelolaan sumberdaya hutan di desa Uwen dan sekitarnya kecamatan Taniwel. www.nfp-unpatti.org/uwen.html. Accessed 30 Mar 2012
14. Tampubolon AP (2008) Study of fuelwood biomass energy policies. Jurnal Analisis Kebijakan Kehutanan 5(1):29–37
15. Sumardjani L, Waluyo SD (2007) Analisa konsumsi kayu nasional. http://www.rimbawan.com/images/stories/makalah/kkn_02mei07_a.pdf. Accessed 28 Mar 2012
16. Budiyanto (2009) Tingkat konsumsi kayu bakar masyarakat hutan desa Hegarmanah Sukabumi. Theses, Dept Forest Management, Forestry faculty, IPB

17. Rachman E, Mahfudz, Kuswandi R (1996) Research on species trial and consumption rate of fuelwood in community forest development program in Wamena. Buletin Penelitian Kehutanan 1:41–50
18. Barnes DF, Krutilla K, Hyde W (2004) The urban household energy transition, energy, poverty, and the environment in the developing World Bank. Resources for the Future, Washington. http://siteresources.worldbank.org/INTGENENERGY/Resources/UrbanEnergy-TransitionV51.pdf. Accessed 28 Jun 2012
19. BPS-Statistic Indonesia (2011) Statistical yearbook of Indonesia 2011. ISSN: 0126–2912 Jakarta BPS Statistic
20. Albalak R, Keeler GJ, Frisancho AR, Haber M (1999) Assessment of PM_{10} concentrations from domestic biomass fuel combustion in two rural Bolivian highland villages. Environ Sci Technol 33:2505–2509
21. Ministry of Forestry (2011) Forestry statistics of Indonesia 2010. ISBN: 979-606-073-6

Characterization of Hydrogen Jet Development in an Argon Atmosphere

Mohd Radzi Abu Mansor and Masahiro Shioji

Abstract Characterization of the hydrogen-jet penetration in an argon atmosphere would first be analyzed in order to find the optimal condition for operating an argon-circulated hydrogen internal combustion engine. In this study, experiments were made in a constant-volume vessel to obtain the hydrogen-jet development data for different conditions of injection pressure, ambient pressure and nozzle diameter in argon and nitrogen atmospheres through observation using high-speed shadow-graph images. Experimental data were then compared with proposed momentum theory calculation in order to describe the development of hydrogen-jet penetration. From the result, $\alpha = 1.25$ was used as an experimental constant. The momentum theory calculation is one of the main important data in helping to set the design parameters for jet penetration and to further established the operational conditions for an argon-circulated hydrogen engine.

Keywords Argon • Argon-circulated hydrogen engine • Constant-volume vessel • Hydrogen • Momentum theory

1 Introduction

Hydrogen is well known as the future energy carrier. It can be used for both renewable and conventional energy systems, such as fuel for internal combustion engine or to create reaction with oxygen in fuel cells to electrochemically produce

---

M.R.A. Mansor
Graduate School of Energy Science, Kyoto University, Kyoto 606-8501, Japan

Faculty of Engineering and Built Environment, The National University of Malaysia, UKM, Bangi 43600, Malaysia

M. Shioji (✉)
Graduate School of Energy Science, Kyoto University, Kyoto 606-8501, Japan
e-mail: shioji@energy.kyoto-u.ac.jp

T. Yao (ed.), *Zero-Carbon Energy Kyoto 2012*, Green Energy and Technology,
DOI 10.1007/978-4-431-54264-3_14,

electrical energy. In an internal combustion engine scenario, the only pollutant to be concerned about from burning hydrogen fuel is NO_x. Replacing normal air with noble-gas-argon and oxygen atmosphere to perform hydrogen combustion would solve the NO_x problem since the reaction only forms water as product. Moreover, the use of high-specific-heat-ratio argon as a working gas dramatically improves the thermal efficiency of an engine's operation [1, 2]. During the compression stroke, the use of the high-specific-heat-ratio working gas argon provide higher pressure and temperatures condition compared to normal air, which makes it possible to use the argon in hydrogen direct-injection compression-ignition engines with low ignitability hydrogen fuel to ignite spontaneously immediately after injection with a shorter ignition delay [3]. Recirculation of rare gas argon need to be considered since the gas is expensive and difficult to carry in large amount. Since the hydrogen-oxygen reaction in noble gas atmosphere forms only water, the reuse of argon can be realized by removing the water element during the exhaust gas recirculation after the combustion process. To date, only a considerable amount of fundamental data for finding the optimal design and operating conditions in argon-circulated hydrogen engine can be obtained [3].

Before further study of combustion process can be made, the characterization of jet development is required to be done for optimization of operational parameter in argon-circulated hydrogen engine. In the case of high-pressure gas fuel directly injected into hot atmosphere, the fuel and oxygen are mixed by entrainment and turbulence in a high-speed unsteady jet rather than by in-cylinder flow, such as swirling or squishing, because the interval between injection and ignition is relatively short [4]. Therefore by controlling the mixture formation in direct injection system greatly affects the combustion process and potentially reductions in exhaust-gas emissions. In this study, the result from hydrogen jet penetration in an argon atmosphere experiment that was done in previous study [5] was compared with momentum theory calculation in order to add information and guideline for hydrogen injection in argon atmosphere. From the calculation, an experimental constant was proposed.

2 Experimental Setup and Procedure

2.1 Constant-Volume Vessel

In this research, experiments were carried out in a constant-volume vessel. The combustion chamber is schematically shown in Fig. 1. The chamber is a circular cylinder, 80 mm in diameter, 30 mm in depth, has approximately 150 cc of volume and is fitted with 95.8 mm diameter, 25 mm thick quartz window for optical access. A solenoid-type gas injector, fitted at the top of the chamber, was used to inject high-pressure hydrogen fuel into the chamber at an injection pressure of up to

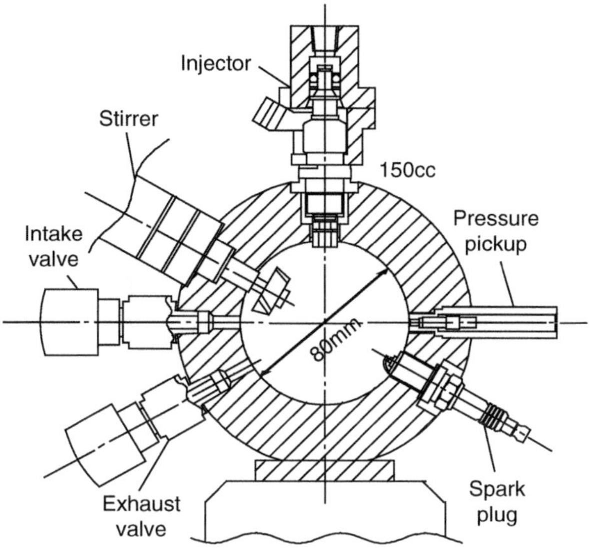

Fig. 1 Cross-section of constant-volume vessel

15 MPa after filling a predetermined ambient pressure of argon or nitrogen. Hydrogen jet penetration are visualized and observed with high-speed shadowgraphs obtained using a digital camera (Photron FASTCAM SA1 675K-M1). The photographs were recorded at a speed of 20,000 fps and an exposure time of 49.3 μs.

2.2 *Hydrogen-Jet Development*

For the purpose of this study, the angle of divergence and penetration length for the hydrogen jet was measured. Figure 2 (right) shows the example of shadowgraph images that were taken during the experiment process. In Fig. 2 (left), the angle of divergence of the jet, β, refers to the projection line on both sides of the hydrogen jet diverging from the injector tip, where the jet penetrated half-way into the constant-volume vessel. In the figure, ρ_f is the density of the fuel, ρ_a is the density of atmosphere and u is the jet velocity at the nozzle exit. The penetration length, x_t, was measured horizontally from the end of the injector tip to the hydrogen jet penetration tip.

By using momentum theory, the amount of injection pressure and the injector orifice diameter can be used to express the fuel penetration length and jet air entrainment speed of mixture formation in cylinder-direct injection [6]. According to momentum theory, when the atmospheric pressure is less than the critical pressure, the fuel penetration length can be calculated using the following equation:

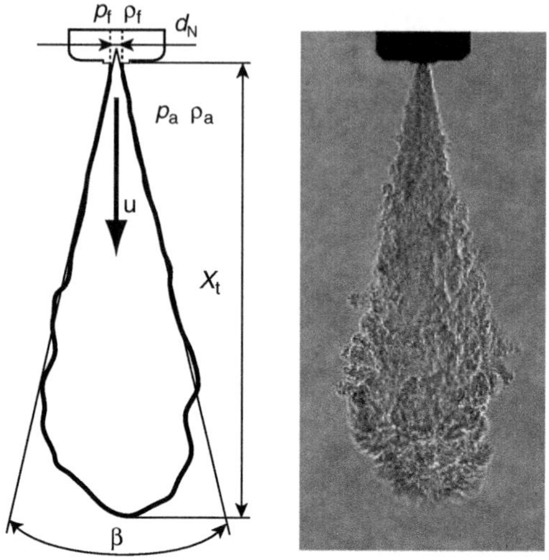

Fig. 2 Jet development assumed in the momentum theory

$$x_t = \alpha\left(\frac{\rho_f}{\rho_a}\right)^{\frac{1}{4}}\left(\frac{d_e ut}{\tan\frac{\beta}{2}}\right)^{\frac{1}{2}} \tag{1}$$

where α is an experimental constant, ρ_{f} is the density of the fuel, ρ_{a} is the density of the atmosphere, u is the jet velocity at the nozzle exit, t is the time from injection start and β is the angle of divergence. In this case, the jet diameter after a notional expansion to ambient pressure is defined as the diameter of the pseudo-nozzle d_{e} [4]:

$$d_e = d_N\sqrt{C_n\frac{p_j}{p_a}\left(\frac{2}{\kappa+1}\right)^{\frac{\kappa+1}{2(\kappa-1)}}} \tag{2}$$

where d_{N} is the injector orifice diameter, C_{n} is the coefficient of discharge, p_{j} is the injection pressure, p_{a} is the ambient pressure and κ is the specific heat ratio. In this study, the experimental constants were considered to have the following values: 1.25; u, sonic speed, 1,320 m/s; and κ for hydrogen: 1.40.

2.3 Experimental Conditions

Table 1 indicates the experimental conditions for the study of the observation of hydrogen jet penetration in an argon atmosphere. The underlined numbers in the table correspond to the standard conditions. Other effects resulting from the changes in these parameters were also observed. It is believed that the data from normal-temperature high-pressure conditions are comparable with high-pressure high-temperature

Table 1 Experimental conditions for hydrogen-jet visualization

Atmosphere	Argon, nitrogen
Injection pressure, p_j (MPa)	4, <u>8</u>, 14
Ambient temperature, T_i (K)	300
Ambient pressure, p_i (MPa)	0.6, <u>1.2</u>, 1.5
Nozzle-orifice diameter, d_N (mm)	0.4, 0.56, <u>0.8</u>

conditions if both were equal in density that lead to the applicability for the obtain data to be use in consideration for operation of argon-circulated hydrogen engine.

3 Results and Discussions

Hydrogen jet development in an argon atmosphere were observed by using the high-speed shadowgraph images, based on which the penetration length, x_t and angle of divergence, β, were obtained. In the following, the penetration length, x_t, were calculated using Eq. (1) and measured value of β, where the results then compared together with the measured one.

3.1 Effects of Different Atmosphere

Development of hydrogen-jet in the argon atmosphere was first examined at standard condition of $p_i = 1.2$ MPa, $T_i = 300$ K, $d_N = 0.8$ mm. Figure 3 shows that the results of the experiments indicated by symbols and the results of calculation indicated by lines. From Fig. 3 both experimental and calculation show nearly identical results for argon. The same result was also observed for an analysis in a nitrogen atmosphere. Because the calculation result was equivalent, momentum theory can be applied for our study on the development of a hydrogen jet with the consideration of the experimental constant α of 1.25.

3.2 Effects of Injection Pressure

Figure 4 shows that the results of the calculation (indicated by lines) and the experiments (indicated by symbols) where the experimental constant α of 1.25 was taken into consideration. From Fig. 4 both experimental and calculation show nearly identical results for $p_j = 8$ MPa and 14 MPa. A slight difference were observed for $p_j = 4$ MPa result. It is believed that the main reason for this difference is mainly due to the accuracy of measurement of angle of divergence β.

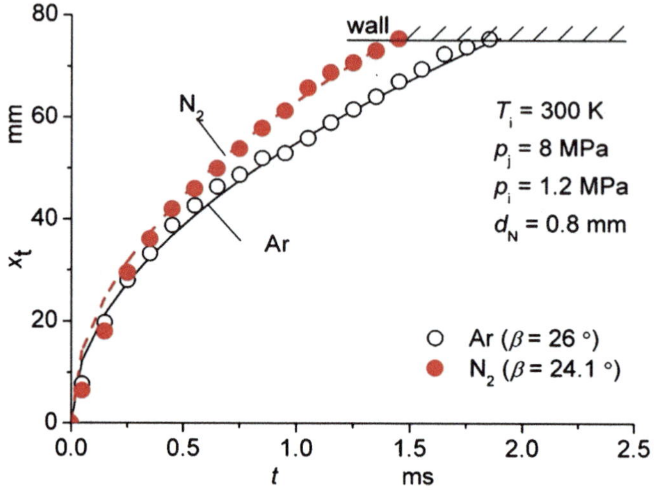

Fig. 3 Calculated results for the penetration length x_t of the hydrogen jet in argon and nitrogen atmospheres

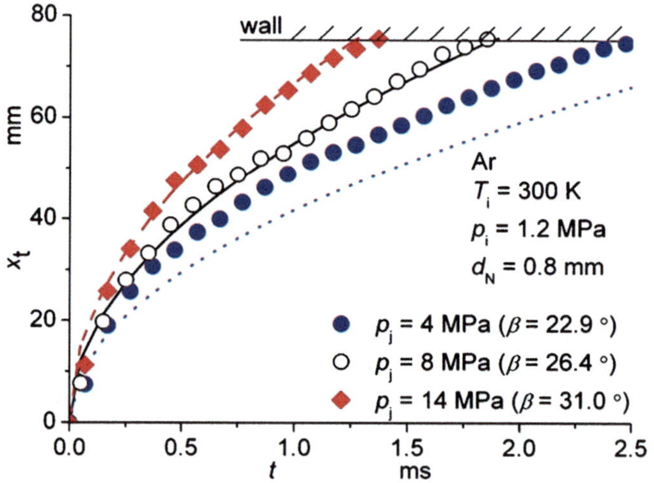

Fig. 4 Calculated results for the penetration length x_t for different injection pressure p_j

3.3 Effects of Ambient Pressure

Figure 5 shows that the results of the calculation (indicated by lines) and the experiments (indicated by symbols) are nearly identical for all ambient pressure conditions even though a slight difference was observed for $p_i = 0.6$ MPa. We also believed if more accuracy measurement of angle of divergence β will result in more accurate result for momentum theory calculation of hydrogen jet penetration.

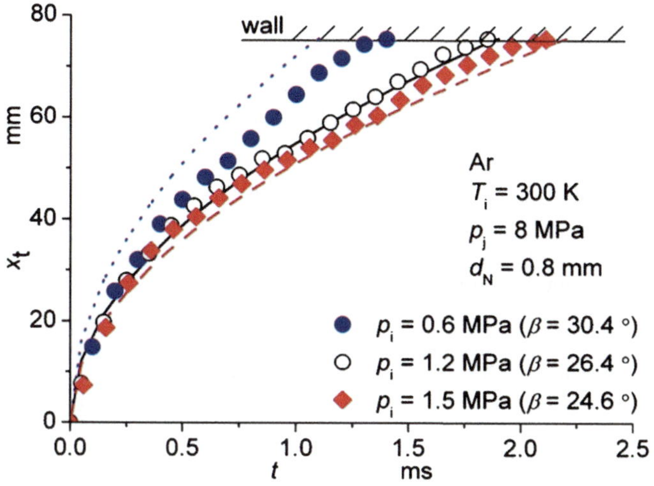

Fig. 5 Calculated results for the penetration length x_t for different ambient pressure p_i

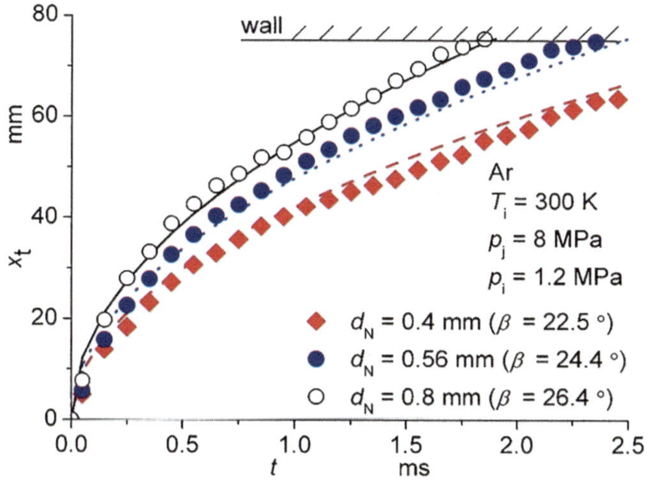

Fig. 6 Calculated results for the penetration length x_t for different nozzle-orifice diameter d_N

3.4 Effects of Nozzle-Orifice Diameter

Next development of hydrogen-jet in the argon atmosphere were also examined at the condition of $p_i = 1.2$ MPa, $T_i = 300$ K, $p_j = 8$ MPa for difference nozzle-orifice diameter, d_N. Figure 6 shows that the results of the calculation (indicated by lines) and the experiments (indicated by symbols) are nearly identical for all nozzle-orifice diameter, d_N.

4 Conclusions

In order to obtain a guideline for hydrogen injection in argon-circulated hydrogen engine, the comparison between hydrogen-jet penetration momentum theory and observation of hydrogen-jet development using high-speed shadowgraph images in a constant-volume vessel filled with argon has been made. It is believed that momentum theory can be applied in our study of the development of hydrogen-jet with the assumption of the experimental constant, being 1.25 since the calculated and experimental results were closely matched.

Acknowledgement The authors were grateful for the support of the Ministry of Education, Culture, Sports, Science and Technology of Japan via "Energy Science in the Age of Global Warming" of Global Center of Excellence (G-COE) program (J-051) and also would like to thank Toyota Motor Company for partly supporting the research grant and providing the gas-injection system used in conducting this study.

References

1. Ikegami M, Miwa K, Shioji M (1982) "A study of hydrogen fuelled compression ignition engines." Int J Hydrogen Energy 7(4):341–353. doi:10.1016/0360-3199(82)90127-6
2. Kuroki R, Kato A, Kamiyama E, Sawada D (2010) "Study of high efficiency zero-emission argon circulated hydrogen engine." SAE Technical Paper 2010-01-0581. doi:10.4271/2010-01-0581
3. Mansor M R A, Nakao S, Nakagami K, Shioji M, Kato A (2012) "Ignition characteristics of hydrogen jets in an argon-oxygen atmosphere." SAE Technical Paper 2012-01-1312. doi:10.4271/2012-01-1312
4. Kawanabe H, Naito Y, Kosaka H, Shioji M (2007) CFD analysis of a high-speed unsteady jet. Heat Transfer Asian Res 36(1):1–12. doi:10.1002/htj.20141
5. Mansor M R A, Nakao S, Nakagami K, Shioji M (2012) "Study on hydrogen-jet development in the argon atmosphere." In: Zero-carbon energy Kyoto 2011, Green energy and technology. Springer, New York, pp 177–184. doi:10.1007/978-4-431-54067-0_21
6. Shioji M, Nakamura K, Miyajima K, Ikegami M (1996) "Study of high-pressure unsteady gas-fuel jets." JSAE Paper Number 9640049

Fabrication of Strontium Titanate Nanofibers via Electrospinning

Lea Macaraig, Surawut Chuangchote, and Takashi Sagawa

Abstract The fabrication of $SrTiO_3$ nanofibers via the post-electrospinning reaction of TiO_2 nanofibers with strontium hydroxide route has led to various heterogeneous forms of strontium-titanium oxides. On the other hand, we prepared homogeneous $SrTiO_3$ nanofibers via the electrospinning of a mixed solution of titanium and strontium precursors. The mixture of precursors was prepared starting with a TiO_2 precursor solution and adding the strontium component by surveying the combination with different types of strontium salts (nitrates, acetates and acetylacetates), solvents (methanol, acetylacetone, and 2-methoxyethanol) with various concentrations of poly(vinylpyrrolidone) as polymer guide and inorganic loadings. It was found that the homogeneous fibers were fabricated from solutions including 2.0 g titanium(IV) butoxide, 1.6 g strontium acetylacetate, 0.6 g acetylacetone, 0.8 g poly(vinylpyrrolidone), and 10 mL acetic acid. Electrospun fibers were characterized via the scanning electron microscopic/energy dispersive spectroscopic observations and the measurement of X-ray diffraction patterns.

Keywords Electrospinning • Nanofiber • Strontium titanate

1 Introduction

The photocatalytic properties of metal oxide semiconductors have been widely studied. Titanium dioxide (TiO_2) has seen the most interest because of its water-splitting capability due to its photostability, availability, low cost, non-toxicity

L. Macaraig (✉) • T. Sagawa
Institute of Advanced Energy, Kyoto University, Gokasho, Uji, Kyoto 611-0011, Japan
e-mail: t-sagawa@iae.kyoto-u.ac.jp

S. Chuangchote
The Joint Graduate School of Energy and Environment, King Mongkut University of Technology Thonburi, 126 Prachauthit Rd., Bangmod, Tungkru, Bangkok 10140, Thailand

T. Yao (ed.), *Zero-Carbon Energy Kyoto 2012*, Green Energy and Technology,
DOI 10.1007/978-4-431-54264-3_15,

and its electronic properties [1, 2]. For water-splitting applications however, the position of the valence band level and the rather large band gap of TiO_2 has inhibited its catalytic performance under visible light. Modification of TiO_2 with the addition of strontium to form strontium titanate ($SrTiO_3$), a perovskite, has been shown to improve the photopotential of TiO_2, allow the production of both hydrogen and oxygen in water-splitting due to the position of the valence and conduction band of $SrTiO_3$, and facilitate easier doping for electronic modification [3]. However, the fabrication of $SrTiO_3$ nanofibers via the hydrothermal reaction of electrospun TiO_2 nanofibers with strontium hydroxide has led to various forms of strontium–titanium heterostructures [4, 5]. In this work, we prepared homogeneous $SrTiO_3$ nanofibers via the electrospinning of a solution, which included both titanium and strontium precursors. The precursor solution was modified from the previously reported and optimized TiO_2 precursor solution [6].

2 Experimental

Precursor solutions were prepared by the addition of equimolar amounts of strontium salt to form a solution with titanium(IV) butoxide (4.0 g; Aldrich; 97%), polyvinylpyrrolidone (PVP; 0.80 g; Aldrich, $M_w = 1{,}300{,}000$), acetylacetone (ACA; 1.2 g; Aldrich) in methanol (10.0 mL; Wako). Strontium nitrate (Wako), strontium acetate (Wako) and strontium acetylacetonate (Aldrich) were tested for solubility with the titanium dioxide precursor solution. Prepared solutions were heated to increase solubilty at 60°C, just below the boiling point of methanol.

Each prepared precursor solution was then loaded into a plastic syringe fitted with a 22-gauge stainless steel needle as the nozzle. The emitting electrode of positive polarity from a Gamma High Voltage Research power supply set at 15 kV was connected to the nozzle, while the grounding electrode was connected to a sheet of aluminum used as the collector plate 15 cm away. The flow rate of the solution was controlled to 1 mL h^{-1}. Complete hydrolysis of the nanofibers was allowed by leaving the nanofibers exposed to atmospheric moisture for 5 h. Precursor solutions prepared at the elevated temperature, were also electrospun at that same temperature by fitting a heated sleeve over the plastic syringe.

The nanofibers were then calcined at 500°C, 600°C, 700°C, and 800°C for 3 h. The morphology and diameter, and crystalline structure of the obtained fibers were characterized by scanning electron microscopy/energy dispersive spectroscopy (SEM/EDS; JEOL JSM-6500FE) and X-ray diffraction spectroscopy (XRD; Rigaku RINT 2100) respectively. The surface areas of the nanofibers were estimated by the Brunauer-Emmett-Teller (BET) technique.

3 Results and Discussion

The precursor solution prepared using strontium nitrate as the strontium salt was not spinnable. Strontium nitrate was not soluble even with the increase of temperature to 60°C. The same could be said with solutions prepared with strontium acetate, although partial dissolution can be observed at the elevated temperature. Strontium acetylacetonate showed partial dissolution at room temperature and formed opaque solutions at 60°C. Electrospun nanofibers from this precursor solution however showed rods and beads (Fig. 1a). EDS analysis of the electrospun fibers showed that rods and beads present in the fiber lower titanium to strontium ratios as compared to the fiber body (Table 1), indicating that the inorganic contents of the precursor solution were not distributed homogenously along the polymer guide. Since beaded fibers are usually formed due to the low viscoelastic force [7] in spinning jets, the viscosity of the precursor solutions was increased by increasing the polymer guide concentration to 10% and 12%. However, the obtained fibers at such concentrations are still with rods and beads (Fig. 1b, c respectively).

Fig. 1 SEM images of $SrTiO_3$ nanofibers fabricated with the use of strontium acetylacetonate salts, methanol as the solvent, with the polymer guide concentration at 8%, 10%, and 12% (**a–c**). All precursor solutions were heated and spun at 60°C. The scale bar is at 5 μm

Table 1 The titanium to strontium ratio of the electrospun fibers as determined by SEM/EDS

Solvent	PVP %	Location	Ti/Sr ratio
Methanol	8	Fiber	1.63
		Rod	0.47
	10	Fiber	2.17
		Rod	0.95
	12	Fiber	1.75
		Rod	1.25
Acetylacetone	8	Fiber	9.22
		Rod	0.75
2-Methoxyethanol	8	Fiber	1.83
		Rod	1.15
Acetic acid	8	Fiber	4.35
		Rod	0.69

Fig. 2 SEM images of $SrTiO_3$ nanofibers fabricated with (**a**) acetylacetone (**b**) 2-methoxyethanol (**c–e**) and acetic acid as the solvent. Precursor solutions (**a–b**) were heated and spun at 60°C. The scale bar of (**a–c**) is at 1 μm and the scale bar of (**d–e**) is at 10 μm

The consistent higher strontium content in the beads and rods (Table 1) even at high viscosities indicate that the beads and rods are not due to low viscoelastic forces but are instead due to incomplete dissolution of the inorganic contents. As such, common organic spinning solvents were surveyed. Acetylacetone, 2-methoxyethanol (Wako) and acetic acid (Wako) were tested as solvents. The polymer guide concentration was fixed at 8% and strontium acetylacetonate was used as the strontium salt.

Similar with the use of methanol as the solvent, opaque solutions were obtained when the precursor solutions were prepared with acetylacetone and 2-methoxyethanol as solvents and were heated at 60°C. However, a clear solution was obtained with the use of acetic acid as the solvent, even without heating. Fibers obtained from precursor solutions of these three solvents still have strontium rich rods and beads (Table 1). The rods and beads in the fibers from the acetic acid precursor solutions however are reduced, and are concentrated in certain areas in the fiber mat as seen in the lower magnification SEM image (Fig. 2d, e).

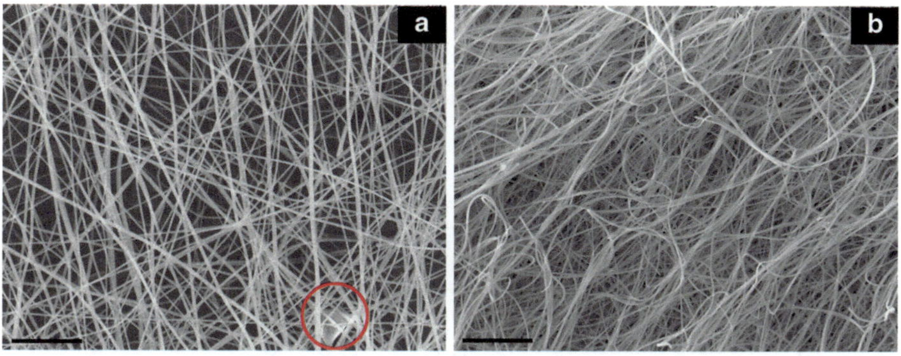

Fig. 3 SEM images of $SrTiO_3$ nanofibers fabricated with the use of strontium acetylacetonate salts, acetic acid as the solvent, with the polymer guide concentration fixed at 8%. Inorganic loading were reduced to 30% (**a**) and 20% (**b**). The scale bar is at 10 μm. The red circle indicates some beads

Since acetic acid solutions are clear and spinnable at room temperature, acetic acid was used as the solvent for all subsequent precursor solution preparations. However, the high titanium to strontium ratio of the fibers (Table 1) and the concentration of rods and beads in the fiber mat indicate that the inorganic content is still not distributed homogeneously. Because of this, the amount of inorganic content loaded onto the precursor solution was reduced to 20% and 30% from 40%. The strontium and titanium stoichiometric ratio were kept constant at 1. Some beads were still present in the electrospun fibers with 30% inorganic loading. The precursor solution with 20% inorganic loading however generated smooth fibers (Fig. 3). Subsequent characterization and calcination work-up were done for fibers fabricated from such precursor solutions.

The inorganic content of the electrospun nanofibers were removed via calcination. Calcination temperatures were varied from 500°C, 600°C, 700°C, and 800°C. The fibers were then characterized for important properties relevant to photocatalytic activity, namely crystallinity, crystallite size and surface area. Fibers obtained after 500°C calcination were beige in color while the higher calcination temperatures resulted in white fibers. Comparing the XRD patterns of the calcined nanofibers to that of a commercial $SrTiO_3$ nanoparticle (CN, Wako), it can be said that the fabricated fibers are indeed $SrTiO_3$ (Fig. 4). Fibers calcined at 500°C are beige in color. This may indicate the presence of organic contents. The low crystallinity of the fibers calcined at 500°C was also confirmed by XRD measurement. Higher crystallinity can be observed for fibers calcined at 700°C and 800°C and are comparable to that of CN. Crystallite size also tends to increase with raising the calcination temperature. But the largest crystallite size of the fibers fabricated is still smaller than that of CN. Small crystallite size is important for photocatalytic activities. The diameter of the nanofibers and surface area however decrease with temperature. Surface areas of the fabricated fibers however are still larger than the commercial nanoparticle except for the fibers calcined at 800°C. while the titanium to strontium ratio is stable across all temperatures and are comparable to that of CN.

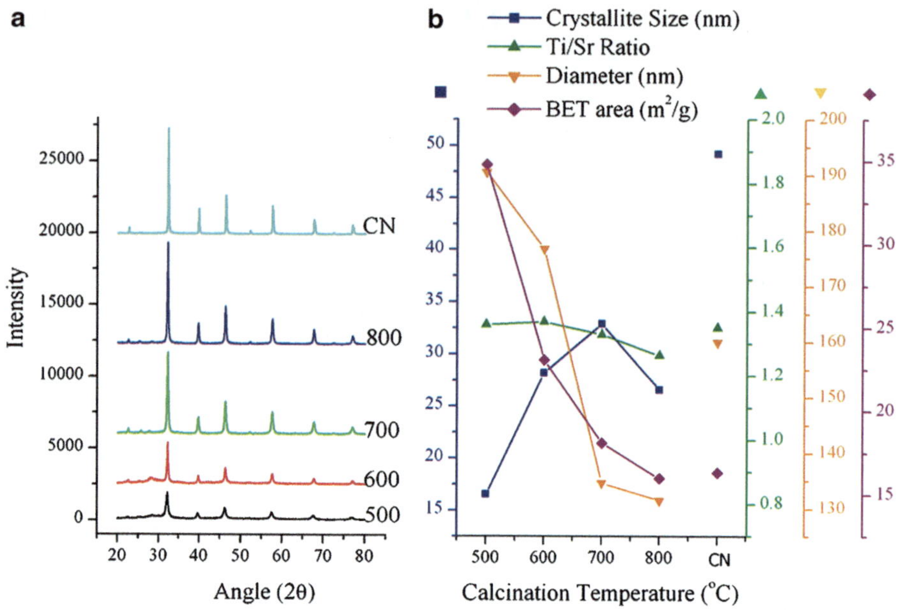

Fig. 4 (**a**) The XRD spectra and (**b**) the properties of the $SrTiO_3$ electrospun nanofibers (calcined at 500°C, 600°C, 700°C, and 800°C) and commercial nanoparticle (CN)

4 Conclusion

Smooth, homogenous $SrTiO_3$ nanofibers were reproducibly fabricated via the electrospinning of clear solutions of 2.0 g TiBu, 1.6 g strontium acetylacetate, 0.6 g acetylacetone, 0.8 g PVP, in 10 mL acetic acid. Acetic acid was found to be the optimum solvent for the production of clear precursor solutions at ambient temperatures. The titanium to strontium ratios of and XRD spectra of the nanofibers are comparable to those of the commercial nanoparticles. Fabricated fibers calcined at 500–800°C (34.8, 23.1, 18.1 and 16.0 $m^2\ g^{-1}$ respectively) have comparable or higher surface areas than commercial $SrTiO_3$ nanoparticles (16.3 $m^2\ g^{-1}$). Electrospun nanofibers also have smaller crystallite size than these nanoparticles (16.5, 28.2, 32.9 and 26.6 nm for fibers calcined at 500–800°C respectively and 49.2 nm for CN). In this regard, it can be said that such fibers can have better photocatalytic potential than the commercial nanoparticle. Also, since the fibers produced were homogenous, metal doping experiments for further improvement of electrical properties can be expected.

Acknowledgment This work was partly supported by the "Energy Science in the Age of Global Warming" of Global Center of Excellence (G-COE) program (J-051) of the Ministry of Education, Culture, Sports, Science and Technology (MEXT) of Japan.

References

1. Fujishima A, Honda K (1972) Electrochemical photolysis of water at a semiconductor electrode. Nature 238(5358):37–38. doi:10.1038/238037a0
2. Ohtani B, Ogawa Y, Nishimoto S-i (1997) Photocatalytic activity of amorphous–anatase mixture of titanium(IV) oxide particles suspended in aqueous solutions. J Phys Chem B 101 (19):3746–3752. doi:10.1021/jp962702
3. Domen K, Kudo A, Onishi T, Kosugi N, Kuroda H (1986) Photocatalytic decomposition of water into hydrogen and oxygen over nickel(II) oxide-strontium titanate ($SrTiO_3$) powder. 1. Structure of the catalysts. J Phys Chem 90(2):292–295. doi:10.1021/j100274a018
4. Cao T, Li Y, Wang C, Shao C, Liu Y (2011) A facile in situ hydrothermal method to $SrTiO_3$/TiO_2 nanofiber heterostructures with high photocatalytic activity. Langmuir 27(6):2946–2952. doi:10.1021/la104195v
5. Macaraig L, Chuangchote S, Sagawa T (2012) Fabrication of $SrTiO_3$ nanofibers for hydrogen production. MRS Proceedings 1408. doi:10.1557/opl.2012.662
6. Chuangchote S, Sagawa T, Yoshikawa S (2009) Electrospinning of poly(vinyl pyrrolidone): effects of solvents on electrospinnability for the fabrication of poly(p-phenylene vinylene) and TiO_2 nanofibers. J Appl Polym Sci 114(5):2777–2791. doi:10.1002/App. 30637
7. Pantano C, Gañán-Calvo AM, Barrero A (1994) Zeroth-order, electrohydrostatic solution for electrospraying in cone-jet mode. J Aerosol Sci 25(6):1065–1077. doi:10.1016/0021-8502(94) 90202-x

A Study on Lithium Recovery from Seawater: Separation of Lithium from Hydrochloric Acid Solutions Containing $CaCl_2$, $MgCl_2$, $MnCl_2$, NaCl, KCl, and LiCl

Namil Um and Tetsuji Hirato

Abstract The adsorption method using manganese oxide is used for lithium recovery from seawater and the product solutions (acidic condition) after adsorption contains other elements such as Ca, Mg, Mn, Na, and K as well as lithium. Thus, Li recovery from hydrochloric acid solutions containing $CaCl_2$, $MgCl_2$, $MnCl_2$, NaCl, KCl, and LiCl by hydrometallurgical process was investigated. A two-stage precipitation was adopted in this process using NaOH to remove Ca, Mg, and Mn and using Na_2CO_3 to separate Li. In the result, Ca, Mg, and Mn could be mostly removed from target solutions by precipitation as $Ca(OH)_2$, $Mg(OH)_2$, and $Mn(OH)_2$ at a range of pH from 11.5 to 12.5 in the first stage. The residual Li that was precipitated as Li_2CO_3 by Na_2CO_3 in the second stage could be concentrated by evaporation at 100°C to make high lithium recovery. The final precipitation of lithium after evaporation produced crystalline lithium carbonates with high purity (99.4%).

Keywords Carbonation • Hydroxide-precipitation • Lithium carbonate • Lithium recovery • Seawater

1 Introduction

An increasing interest has developed in lithium resources for the wide application in large-capacity rechargeable batteries, light aircraft alloys, and nuclear fusion fuels and so on [1]. At this point in time, seawater is considered as a vast source [2, 3] although the concentration of lithium is very low, and the adsorption method using manganese oxide is widely used for lithium recovery from seawater [4–6]. In this

N. Um • T. Hirato (✉)
Department of Energy Science and Technology, Kyoto University, Yoshida-honmachi, Sakyo-ku, Kyoto 606-8501, Japan
e-mail: hiratou@energy.kyoto-u.ac.jp

T. Yao (ed.), *Zero-Carbon Energy Kyoto 2012*, Green Energy and Technology, DOI 10.1007/978-4-431-54264-3_16,

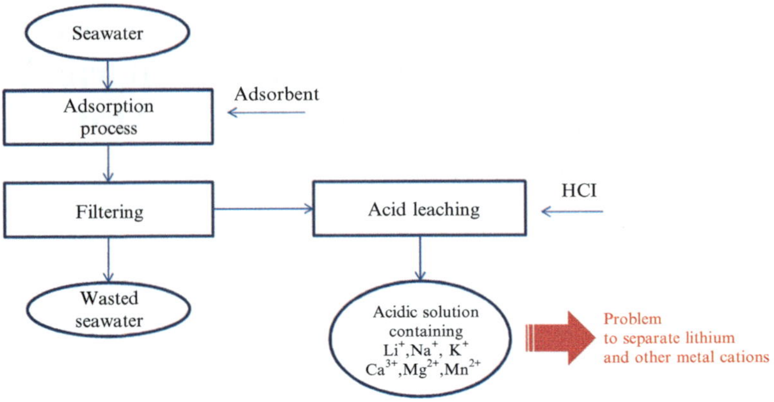

Fig. 1 Schematic flow diagram of lithium recovery from seawater using adsorption method

method (Fig. 1), however, the product solutions after adsorption process contain other elements as well as lithium, thus it is needed to separate lithium from product solutions.

In this study, Li recovery from the hydrochloric acid solutions containing $CaCl_2$, $MgCl_2$, $MnCl_2$, NaCl, KCl, and LiCl was investigated. Firstly the phase equilibrium of $CaCl_2$–$MgCl_2$–$MnCl_2$–NaOH–H_2O system was interpreted with OLI-Systems software which implements the chemical modeling of aqueous electrolyte solution; the precipitation behaviors of $Ca(OH)_2$, $Mg(OH)_2$, and $Mn(OH)_2$ in $CaCl_2$–$MgCl_2$–$MnCl_2$–NaOH–H_2O solutions at various temperatures and pH values were estimated by chemical modeling. The obtained simulation results of precipitation behaviors were compared with available experimental data and good agreement was found. According to the below results it is possible to remove Ca, Mg, and Mn by hydroxide precipitation using NaOH and to separate Li from residue solutions by Li_2CO_3 precipitation using Na_2CO_3 after hydroxide precipitation.

2 Experimental Methods and Chemical Modeling

2.1 Materials and Experiment Methods

In this study lithium chloride (LiCl), sodium chloride (NaCl), potassium chloride (KCl), calcium chloride ($CaCl_2$), magnesium chloride (MgCl), and manganese chloride ($MnCl_2$) from Wako Pure Chemical Industries were used and particle size of all reagents was below 5 μm. All experiments of Ca, Mg and Mn removal and Li recovery from hydrochloric acid solutions were carried out by using a batch glass reactor of 500 mL.

0.0015 mol $CaCl_2$, 0.0015 mol $MgCl_2$, 0.0015 mol $MnCl_2$, 0.0015 mol NaCl, 0.0015 mol KCl, and 0.025 mol LiCl were added to 150 mL of 0.5 mol/dm^3 hydrochloric acid solutions. This target solutions was heated and kept at the temperature of 90°C under atmospheric pressure. And it was stirred by magnetic stirrer at 650 rpm. Also, pH was adjusted to the required values from 7 to 14 by NaOH. A pH meter (Horiba, F-52) was used to check the hydrogen ion concentration and the concentrations of Ca^{2+}, Mg^{2+}, and Mn^{2+} at each pH values were determined by ICP-AES (Seiko Instruments Inc., SPS7800).

The residual 0.025 mol Li in leachate after removal of hydroxide precipitation was precipitated as Li_2CO_3 using 0.0125 mol Na_2CO_3. And then evaporation tests were conducted to decrease solution volume from 150 mL to 50 mL by continuous boiling (100°C) and concentrated Li_2CO_3 was filtered from the solution on a 0.45 μm pore size of membrane filter using a pressure filtration unit. The crystal structure of precipitate was determined by X-ray diffractometer (PANalytical, X'pert PRO).

2.2 *Chemical Modeling*

The chemical modeling for electrolytes can be useful in speciation-based models, which involve multiple chemical equilibria and have an advantage at moderate temperatures and concentrations (up to the saturation point). Therefore, a speciation-based model can be used to depict an aqueous processing system at high temperature. In this study, OLI-systems® software that implements the chemical modeling of aqueous electrolyte solution was used to fit the chemical equilibria of Ca, Mg, and Mn in alkali solution. Especially, the precipitation behavior of Ca $(OH)_2$, $Mg(OH)_2$, and $Mn(OH)_2$ in $CaCl_2$–$MgCl_2$–$MnCl_2$–NaOH–H_2O solutions under different pH values at 90°C could be estimated.

3 Results and Discussions

This study investigates the conditions for removal of Ca, Mg, and Mn through $Ca(OH)_2$, $Mg(OH)_2$, and $Mn(OH)_2$ precipitation according to NaOH injection into target solutions and for recovery of Li through Li_2CO_3 precipitation using Na_2CO_3 and concentration by evaporation to achieve high lithium recovery. Figure 2 outlines the total steps involved in the removal of Ca, Mg, and Mn by first stage process and the recovery of Li by second stage process.

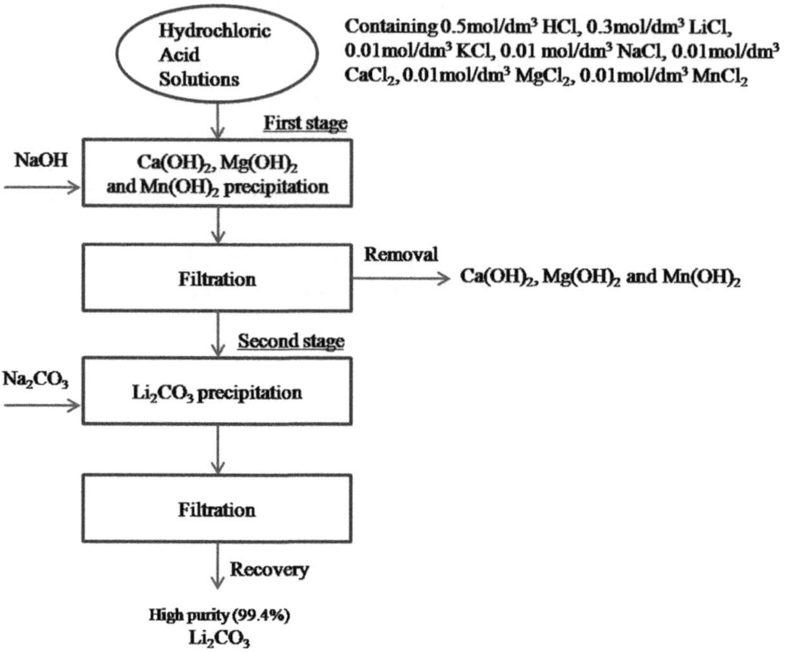

Fig. 2 Total process of Li recovery from hydrochloric acid solutions containing $CaCl_2$, $MgCl_2$, $MnCl_2$, NaCl, KCl, and LiCl

3.1 Removal of Calcium, Magnesium and Manganese

OLI-systems® software, which implements the chemical modeling, was first used to predict the conditions for selectively precipitating Ca, Mg, and Mn as different pH values ranging from 7 to 14 at 90°C. A typical graphs of the results obtained from chemical modeling against experimental data is shown in Fig. 3. The chemical model is in good agreement with experimental data. These results shows that all Ca and Mg could be precipitated as $Ca(OH)_2$ and $Mg(OH)_2$ at a pH > 9.0 and pH > 12.0, respectively. Whereas the precipitation of all $Mn(OH)_2$ takes place over a range of pH from 8.6 to 12.5. As a result, the removal of Ca, Mg, and Mn is controlled by these precipitates during NaOH addition (or increasing pH). Li, Na, and K is stable as soluble Li^+, Na^+, and K^+ ions over a wide range of pHs. Therefore, first stage based on precipitation experiment and chemical modeling showed that more than 90% of 0.0015 mol Ca, 0.0015 mol Mg, and 0.0015 mol Mn could be removed selectively in the first stage using NaOH.

3.2 Lithium Recovery

After Ca, Mg, and Mn were removed the leachate contained Li, K, and Na. Following the second stage process, Li could be precipitated as Li_2CO_3 by

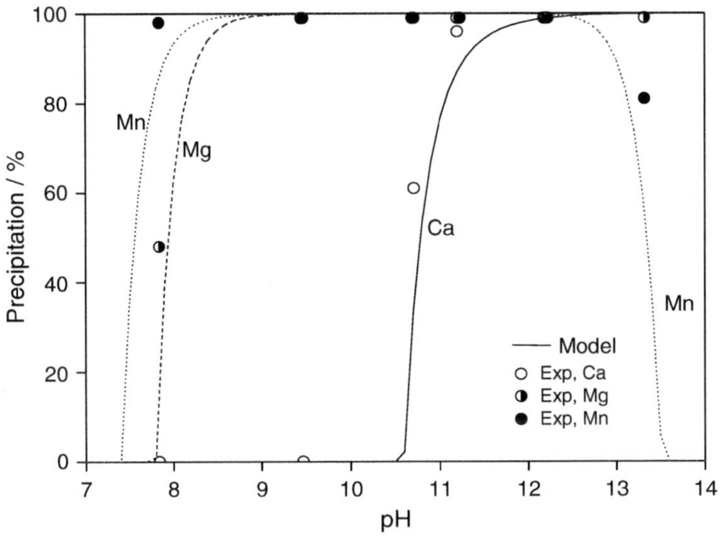

Fig. 3 Precipitation behavior of $Ca(OH)_2$, $Mg(OH)_2$, and $Mn(OH)_2$ at different pH values in $CaCl_2$–$MgCl_2$–$MnCl_2$–NaOH–H_2O solutions (90°C reaction temperature and 0.01 mol/dm^3 $CaCl_2$, $MgCl_2$, and $MnCl_2$ initial amount)

Na_2CO_3 and concentrated by evaporation at 100°C. These operations were performed using the leachate obtained from the first stage process in which precipitated $Ca(OH)_2$, $Mg(OH)_2$, and $Mn(OH)_2$ were removed under pH 11.6 and 90°C. As the result, the recovery of Li in the second stage process was lower than 60% of 0.025 mol initial Li; the possible precipitation of Li_2CO_3 in alkali solutions depends on the solubility limits at different pH values and the composition of target solutions, thus it is needed to investigate how to obtain higher Li recovery than expected, in the future plan. The final precipitation product was a well crystalline material as confirmed by its XRD pattern and was shown in Fig. 4. The high purity product of 0.0145 mol Li_2CO_3 (99.4% purity with <0.6% of Ca, Mg, Mn, K, and Na) could be recovered.

4 Conclusions

A hydrometallurgical process was developed to separate Li from the hydrochloric acid solutions containing $CaCl_2$, $MgCl_2$, $MnCl_2$, NaCl, KCl, and LiCl. A two-stage precipitation was employed to remove Ca, Mg, and Mn using NaOH and to separate residual Li using Na_2CO_3. Precipitated $Ca(OH)_2$, $Mg(OH)_2$, and $Mn(OH)_2$ in the first stage process needs to be removed to produce high purity Li product.

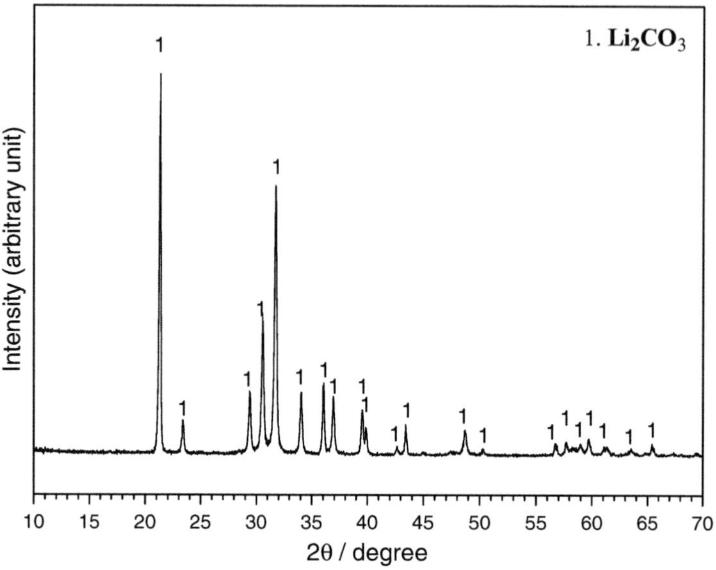

Fig. 4 XRD pattern of the precipitate after lithium carbonation

The residual Li can be formed as Li_2CO_3 in the second stage process and concentrated Li_2CO_3 by evaporation at 100°C lead to obtain high Li recovery. A high purity (99.4%) and well crystalline Li_2CO_3 could then be produced from this process.

Acknowledgement This work was partly supported by an "Energy Science in the Age of Global Warming" of Global Center of Excellence (G-COE) program (J-051) of the Ministry of Education, Culture, Sports, Science and Technology of Japan.

References

1. Ebensperger A, Maxwell P, Moscoso C (2005) The lithium industry: its recent evolution and future prospects. Resour Policy 30:218–231
2. Riley JP, Tongudai M (1964) The lithium content of sea water. Deep Sea Res 11:563–568
3. Angino EE, Billings GK (1966) Lithium content of sea water by atomic absorption spectrometry. Geochim Cosmochim Acta 30:153–158
4. Wang L, Meng CG, Ma W (2009) Study on Li^+ uptake by lithium ion-sieve via the pH technique. Colloids Surf A 334:34–39
5. Chung KS, Lee JC, Kim WK, Kim SB, Cho KY (2008) Inorganic adsorbent containing polymeric membrane reservoir for the recovery of lithium from seawater. J Membr Sci 325:503–508
6. Zhang QH, Sun S, Li S, Jiang H, Yu JG (2007) Adsorption of lithium ions on novel nanocrystal MnO_2. Chem Eng Sci 62:4869–4874

Synthesis of Silver Nano-Wires Without Seeds by the Polyol Process

Duck-Hyun Song and Tetsuji Hirato

Abstract Nano-scale materials have generated significantly scientific interest due to their unique optical, electrical and magnetic properties. The properties of nanomaterial are strongly dependent on the size and morphology of the particles. Therefore, it is very important to finely control the size and morphology of the nano-materials. In this work, one-dimensional structured silver nano-wires were synthesized on slide glass by polyol process without seeds. It was found that silver nano-wires were formed in presence of polyvinylpyrrolidone as capping agents. The obtained silver nano-wires were well-high mono-dispersed. Morphology of these nano-wires was changed by controlling the reaction temperature.

Keywords Chemical reduction method • One-dimensional structure • Polyol process • Silver nano-rod • Silver nano-wire

1 Introduction

It is well-known that nano-size particles have been applied for many applications such as optical materials, electronic materials, absorbents, catalysts, etc. owing to their unique electrical, optical, physical, and chemical properties [1]. The property of nano-materials is influenced by size and morphology [2]. Particularly, 1-dimensonal structure materials such as rods, wires, tubes, and belts are anticipated important part of nano-scale electronic devices as both interconnectors and active components because of their special electrical and physical properties [3].

Silver has the highest electrical conductivity (15.87 nΩ m at 20°C) and thermal conductivity (429 W m^{-1} K^{-1}) of all the metals, by virtue of which silver nano-wires

D.-H. Song • T. Hirato (✉)
Department of Energy Science and Technology, Kyoto University, Yoshida-honmachi, Sakyo-ku, Kyoto 606-8501, Japan
e-mail: hiratou@energy.kyoto-u.ac.jp

T. Yao (ed.), *Zero-Carbon Energy Kyoto 2012*, Green Energy and Technology, DOI 10.1007/978-4-431-54264-3_17,

have been used in many applications, such as optical polarizers [4], photonic crystals [5], and catalysts [6]. In recent year, various methods have been used for synthesizing silver nano-wires such as the polyol process [7, 8], wet chemical synthesis [9, 10], the hydrothermal method [11–13], electrochemical techniques [14, 15], porous materials templates [16, 17], and solvothermal method [18].

This article describes the fabrication of silver nano-wire on slide glass by polyol process without Pt (or Ag) seed and discusses about effect of reaction temperatures on the morphologies. For synthesis of silver nano-wires without seeds, we controlled dropping speed of silver nitrate solution.

2 Materials and Methods

2.1 *Materials*

Lithium chloride (LiCl, 99%) was purchased from Wako Pure Chemical Industries, Ltd. Silver nitrate ($AgNO_3$, >99%), polyvinylpyrrolidone (PVP, molecule weight $\approx$ 55,000), and ethylene glycol (E.G., 99.8%) was purchased from Sigma-Aldrich.

2.2 *Methods*

Silver nano-wires were synthesized on slide glass by polyol process as follows. The desired amount of PVP (0.1 M) and LiCl (2 mM) was dissolved in 10 mL of E.G. at the vial. After 10 min stirring, ultrasonically cleaned slide glass (1 mm $\times$ 2 mm) was placed in the solution. And then the solution was heated at various reaction temperatures and thermally stabilized in an oil bath. After the temperature was stabilized, 5 mL of $AgNO_3$ solution (0.1 M, in E.G.) was titrated for 30 min at the rate of 10 mL/h because of this speed could be enough time to nucleate of silver atoms titrated in the initial stages. The solution temperature was maintained during the titration. After silver source was supplied, the solution was heated for an additional 60 min to ensure that the growth is complete. Soft stirring was maintained during the whole of the entire process. And the entire process of experiment was exposed to sun light. Because silver ion titrated in the initial stages could grow up to decahedra seeds upon light exposure [19]. After reaction, solution was cooled-down at room temperature. The slide glass was cleaned three times with ethanol to remove solvent, PVP, and other impurities. And then the slide glass dried at 90°C for 12 h. The morphology and size of the synthesized particles was observed using scanning electron microscopy (SEM, JEOL JSM-6510LV).

3 Results and Discussion

Mostly, synthesis of metal nano-particles through reducing intermediate phase by polyol process can be considered as follows. In the initial state, liquid polyol dissolved the intermediate phase. And then dissolved intermediate phase are reduced by polyol. In the solution, spontaneous nucleation and growth of metal particles are occurred [20]. The possible chemical reactions can be schematized as follows:

$$HOCH_2CH_2OH \rightarrow CH_3CHO + H_2O \tag{1}$$

$$CH_3CHO + AgNO_3 + H_2O \rightarrow Ag + HNO_3 + CH_3COOH \tag{2}$$

In the first step, ethylene glycol can be reversibly converted into an acetaldehyde through dehydrogenation reaction (Eq. 1) [21, 22]. In the second step, $AgNO_3$ was added titrate-wise to the reaction system, Ag nano-particles were formed by reducing $AgNO_3$ with acetaldehyde. Silver atoms nucleated through the homogeneous nucleation process and silver nano-particles were well-dispersed by virtue of the presence of PVP. A polymeric surfactant (PVP) could chemically adsorb onto the silver surfaces through O–Ag bonding [23]. When the solution containing silver nano-particles was heated continuously at high temperatures, small silver nano-particles dissolved, and redeposited onto larger silver nano-particles via a process known as Ostwald ripening [24, 25] because of larger particles are more energetically favored than smaller particles. At this time, silver could grow to the intrinsic crystal due to the existence of PVP which were bonded with silver surfaces and the intrinsic crystal growth preference of silver.

Temperature was important factor to synthesize silver nano-wire. Figure 1 shows the SEM images of silver particles synthesized at various reaction temperatures for 1 h. When the reaction temperature was 120°C (Fig. 1a), silver nano-wires were not detected. The silver atoms grew to the diverse sized micro-particles with 2-D plate and some cuboctahedron. It seemed that the relatively low reaction temperature could not provide enough energy required for thc activation of specific faces for the anisotropic growth of nano-wires. Both the dissolution of small silver nano-particles and the diffusion of silver atoms on the surface of silver nano-wires should require a relatively high temperature. Conversely, silver particles synthesized at 200°C, seem micro-rods that were assembled thin nano-wires (Fig. 1c). Because, relatively high reaction temperature, PVP bonding structures, which were bonded with silver surfaces, easily slackened and PVP could not play capping agent and protective agent. Consequently thin silver nano-wires were combined to the silver micro-rods, and that could be caused by the influence of gravity between the thin silver nano-wires. The side surface and edge of silver micro-rods showed the evidence of the fusion of thin silver wires (Fig. 1d). Figure 1b shows the silver nano-wires synthesized at 160°C. The synthesized nano-wires had a mean diameter of 38–46 nm, and an average length of 15–40 μm as

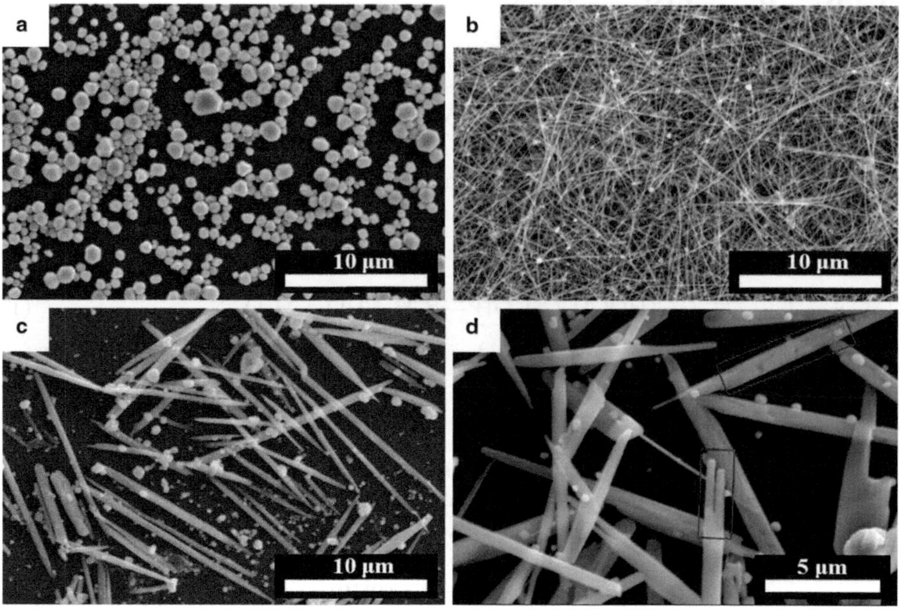

Fig. 1 SEM images of silver nano-particles synthesized by polyol process at various reaction temperatures for 1 h: (**a**) 120°C, (**b**) 160°C, (**c**) and (**d**) 200°C

determined by scanning electron microscope (SEM). These silver nano-wires were relatively uniform in shape and size, and present in high yield (more than 90% relative to other structures of silver). This reaction temperature could provide enough thermal energy for anisotropic growth, and PVP were stably bonded with silver surfaces as capping agent.

4 Conclusions

In summary, various morphologies of silver particles were successfully synthesized by polyol process without any seeds. The morphology of silver was controlled by regulating the reaction temperatures. Uniform silver nano-wires were successfully synthesized at 160°C. At relatively low reaction temperature, 2-D plate and cuboctahedron of silver was synthesized. On the contrary, micro-rods of silver were synthesized at relatively high reaction temperature.

Acknowledgement This work was partly supported by an "Energy Science in the Age of Global Warming" of Global Center of Excellence (G-COE) program (J-051) of the Ministry of Education, Culture, Sports, Science and Technology of Japan.

References

1. Alivisatos AP (1996) Semiconductor clusters, nanocrystals, and quantum dots. Science 271–5251:933–937
2. Volokitin Y, Sinzig J, de Jongh LJ, Schmid G, Vargaftik MN, Moiseev II (2000) Quantum-size effects in the thermodynamic properties of metallic nanoparticles. Nature 384–6610:621–623
3. Sun Y, Xia Y (2002) Large-scale synthesis of uniform silver nanowires through a soft, self-seeding, polyol process. Adv Mater 14–11:833–837
4. Pang YT, Meng GW, Fang Q, Zhang LD (2003) Silver nanowire array infrared polarizers. Nanotechnology 14:20–24
5. Hu X, Chan CT (2004) Photonic crystals with silver nanowires as a near-infrared superlens. Appl Phys Lett 85:1520–1522
6. Bi Y, Lu G (2008) Morphology-controlled preparation of silver nanocrystals and their application in catalysis. Chem Lett 37:514–515
7. Reyes-Gasga J, Elechiguerra JL, Liu C, Camacho-Bragado A, Montejano-Carrizales JM, Yacaman MJ (2006) On the structure of nanorods and nanowires with pentagonal cross-sections. J Cryst Growth 286:162–172
8. Skrabalak SE, Wiley BJ, Kim M, Formo EV, Xia YN (2008) On the polyol synthesis of silver nanostructures: glycolaldehyde as a reducing agent. Nano Lett 8:2077–2081
9. Caswell KK, Bender CM, Murphy CJ (2003) Seedless, surfactantless wet chemical synthesis of silver nanowires. Nano Lett 3:667–669
10. Mdluli PS, Revaprasadu N (2009) An improved N, N-dimethylformamide and polyvinyl pyrrolidone approach for the synthesis of long silver nanowires. J Alloys Compd 469:519–522
11. Wang ZH, Liu JW, Chen XG, Wan JX, Qian YT (2005) A simple hydrothermal route to large-scale synthesis of uniform silver nanowires. Chem Eur J 11:160–163
12. Xu J, Hu J, Peng CJ, Liu HL, Hu Y (2006) A simple approach to the synthesis of silver nanowires by hydrothermal process in the presence of Gemini surfactant. J Colloid Interface Sci 298:689–693
13. Yang Z, Qian H, Chen H, Anker JN (2010) One-pot hydrothermal synthesis of silver nanowires via citrate reduction. J Colloid Interface Sci 352:285–291
14. Mazur M (2004) Electrochemically prepared silver nanoflakes and nanowires. J Electrochem Soc 6:400–403
15. Huang LM, Wang HT, Wang ZB, Mitra A, Bozhilov KN, Yan YS (2002) Nanowire arrays electrodeposited from liquid crystalline phases. Adv Mater 14:61–64
16. Berchmans S, Nirmal RG, Prabaharan G, Madhu S, Yegnaraman V (2006) Templated synthesis of silver nanowires based on the layer-by-layer assembly of silver with dithiodipropionic acid molecules as spacers. J Colloid Interface Sci 303:604–610
17. Kong LB, Lu M, Li MK, Li HL, Guo XY (2003) Branched silver nanowires obtained in porous anodic aluminum oxide template. J Mater Sci Lett 22:701–702
18. Chen D, Qiao X, Qiu X, Chen J, Jiang R (2010) Convenient synthesis of silver nanowires with adjustable diameters via a solvothermal method. J Colloid Interface Sci 344:286–291
19. Pietrobon B, Kitaev V (2008) Photochemical synthesis of monodisperse size-controlled silver decahedral nanoparticles and their remarkable optical properties. Chem Mater 20:5186–5190
20. Fievet F, Lagier JP, Blin B, Beaudoin B, Figlarz M (1989) Homogeneousand heterogeneous nucleations in the polyol process for the preparation of micron and submicron size metal particles. Solid State Ionics 32–33(1):198–205
21. Sun Y, Yin Y, Mayers BT, Herricks T, Xia Y (2002) Uniform silver nanowires synthesis by reducing $agno_3$ with ethylene glycol in the presence of seeds and poly(vinyl pyrrolidone). Chem Mater 14:4736–4745
22. Smith WB (2002) Ethylene glycol to acetaldehyde-dehydration or a concerted mechanism. Tetrahedron 58:2091–2094

23. Huang HH, Ni XP, Loy GL, Chew CH, Tan KL, Loh FC, Deng JF, Xu GQ (1996) Photochemical formation of silver nanoparticles in poly(N-vinylpyrrolidone). Langmuir 12:909–912
24. Roosen AR, Carter WC (1998) Simulations of microstructural evolution: anisotropic growth and coarsening. Physica A (Amsterdam) 261:232–247
25. Matijevic E (1993) Preparation and properties of uniform size colloids. Chem Mater 5:412–426

Relaxation Analysis of Li Inserted γ-Fe_2O_3 at Various Rates

Seungwon Park, Kaoru Takasu, and Takeshi Yao

Abstract Lithium was electrochemically inserted into γ-Fe_2O_3 with different current densities and its crystal structure was analyzed by X-ray Rietveld method. Iron migrated from 8*a* site to 16*c* site in the process of lithium insertion and from 16*c* site to 8*a* site gradually at relaxation process after the lithium insertion. It was considered that lithium was inserted at 8*a* site during the lithium insertion process and migrated to 16*c* site with relaxation time. Iron occupancy of 8*a* site just after lithium insertion increased with the decrease of insertion rate. It was considered that lithium migrated from 8*a* site to 16*c* site even during lithium insertion process at low insertion rate. These results were very well consistent with the site preference that lithium prefers 8*a* site to occupy kinetically and prefers 16*c* site thermodynamically.

Keywords γ-Fe_2O_3 • Lithium ion secondary battery • Relaxation analysis • Spinel-type structure • The Rietveld method

1 Introduction

In recent years, we have been conducting the relaxation analysis of electrode materials such as γ-Fe_2O_3 [1, 2], $LiFePO_4$ [3] and $LiMn_2O_4$ [4] after termination of lithium insertion. The relaxation analysis can make clear the transition process from kinetic state at usage to equilibrium state at rest.

In the previous study, we analyzed the phase change of $LiFePO_4$ with olivine-type structure at relaxation state [3]. It was observed that the amount of $LiFePO_4$ decreased and that of $FePO_4$ increased after termination of lithium insertion. It was

S. Park • K. Takasu • T. Yao (✉)
Department of Fundamental Energy Science, Graduate School of Energy Science, Kyoto University, Yoshida-Honmachi, Sakyo-ku, Kyoto 606-8501, Japan
e-mail: yao@energy.kyoto-u.ac.jp

T. Yao (ed.), *Zero-Carbon Energy Kyoto 2012*, Green Energy and Technology,
DOI 10.1007/978-4-431-54264-3_18,

considered that $LiFePO_4$ phase including lithium defects favorable lithium diffusion formed in the process of lithium insertion and that the defective $LiFePO_4$ separated to $LiFePO_4$ without defects and $FePO_4$ at relaxation process after the lithium insertion.

Also, we conducted the phase change analysis of $LiMn_2O_4$ with spinel structure at relaxation process [4]. It was found that two phases, Li-rich phase and Li-lean phase, coexist [5] and that the amount of Li-lean phase decreased and that of Li-rich phase increased with relaxation time after lithium insertion. It was considered that Li-lean phase has more defects than Li-rich phase, and that Li-lean phase is kinetically favorable for lithium to diffuse. From this reason, the excessive Li-lean phase formed in the process of lithium insertion. And, because Li-rich phase is thermodynamically more stable than Li-lean phase, Li-lean phase decreased and Li-rich phase increased at relaxation process.

γ-Fe_2O_3 has a defective spinel structure and has been attracting as an electrode material for lithium ion secondary battery due to its low cost and low environmental impact [1, 2, 6, 7]. Previously, we analyzed the crystal structure change of γ-Fe_2O_3 after lithium insertion process [1, 2]. At lithium insertion process, the iron occupancy of 8*a* site decreased and that of 16*c* site increased. After the lithium insertion, the iron occupancy of 8*a* site increased and that of 16*c* site decreased gradually with relaxation time. The iron occupancy of 8*b* site was almost zero and that of 16*d* site was almost unchanged all over the process. At lithium insertion process, it was considered that lithium was inserted at 8*a* site and iron migrated from 8*a* site to 16*c* site. After the lithium insertion, it was considered that lithium migrated from 8*a* site to 16*c* site and iron moved from 16*c* site to 8*a* site. From this point of view, we concluded that lithium prefers 8*a* site to occupy kinetically and prefers 16*c* site thermodynamically.

In the present study, in order to understand the electrode behavior more clearly, we inserted lithium into γ-Fe_2O_3 electrochemically with different current densities and analyzed the crystal structure of lithium inserted γ-Fe_2O_3 at relaxation time by means of X-ray diffraction and the Rietveld method. From the results, we discussed the relaxation process on γ-Fe_2O_3 at both kinetic state and thermodynamic state after termination of lithium insertion.

2 Experimental

2.1 *Sample Preparation*

γ-Fe_2O_3 was prepared by aqueous solution method. $FeCl_2 \cdot 4H_2O$ (Wako Chemical) was added to a buffer solution of pH 6.2. The solution was stirred for 30 min at 25°C with producing O_2 gas. The buffer solution was prepared beforehand from 192 mL of 0.1 M potassium acetate (Hayashi Pure Chemical) and 6 mL of 0.1 M acetic acid (Wako Chemical). By this treatment, γ-FeOOH was precipitated in the solution.

The precipitate was collected by suction filtration using a 0.1 μm polytetrafluoroethylene (PTFE) filter and was then treated thermally at 70°C for 24 h under ambient air and subsequently at 200°C for 72 h in a vacuum.

Lithium was electrochemically inserted into the synthesized γ-Fe_2O_3 using an Ar-sealed three-electrode glass beaker cell. For the working electrode, the γ-Fe_2O_3 was mixed with AB (Acetylene Black, surface area: 133 m^2 g^{-1}, Denkikagaku Kogyo Corp., Ltd) as a supplemental conductor and PTFE powder as an adhesive agent in a weight ratio of 0.70:0.30:0.05. The mixture was ground, spread and pressed onto a nickel mesh as a current collector. Lithium metal was used for the counter and reference electrodes. One molar ethylene carbonate and a 1,2-dimethoxyethane solution (1:1, v/v) of lithium perchlorate ($LiClO_4$ EC/DME, Kishida Chemical Corp., Ltd) was used for the electrolyte. The electrode fabrication and the cell assembly were carried out under argon gas system.

The working electrode was discharged from the natural potential of approximately 3 V (vs. Li/Li^+) for lithium insertion. The same amount of lithium was inserted with different current densities of 0.1, 0.01 and 0.001 Ag^{-1} corresponding to the lithium insertion of 0.6, 0.06, and 0.006 mol into γ-Fe_2O_3 for an hour, respectively. The amount of inserted lithium was calculated by integrating the current, which correspond to $x = 1.5$ in terms of $Li_xFe_2O_3$. When the required electrochemical condition was attained, the circuit was opened and the working electrode was removed from the cell immediately under argon atmosphere to avoid the local cell reaction between the electrode material and the current collector or the supplemental conductor. Local cell reaction should be avoided because it is fully different from relaxation and make the observation of relaxation process impossible.

2.2 *Structural Characterization*

The electrode at relaxation was set in a sealed holder (2391A201, Rigaku Corp., Ltd.) under argon gas system. The sealed holder was set to the diffractometer (UltimaIV, Rigaku Corp., Ltd.) for X-ray diffraction (XRD) measurement. XRD patterns were continuously measured from 25° to 135° in 2θ at a rate of 2° per min with 0.04° step width by using CuKα radiation. Tube voltage and current were set to 40 kV and 40 mA, respectively. Continuous ten patterns were summed to make noise decrease and the Rietveld analysis precise. Elapsed time after lithium insertion was represented by the mean.

Observed XRD patterns were analyzed by the Rietveld method using RIEVEC program coded by T. Yao [1–3, 7–9]. The crystal structure was represented by the space group of $Fd3m$, and the occupancies of iron at 8*a*, 8*b*, 16*c* and 16*d* sites were analyzed. It was assumed that the atomicity of iron was constant both before and after lithium insertion. The effect of reflection from lithium was ignored, and peaks derived from the AB and the nickel collector included in the XRD patterns were evaluated as background.

3 Results and Discussion

XRD profiles of the samples for every relaxation time were well-fitted with the calculated patterns by the Rietveld method. Figure 1 shows fitting results of the sample before lithium insertion and the sample after lithium insertion at 0.01 Ag^{-1} followed by the relaxation for 5 h. The resulting reliability values of R_{WP}, R_F and R_B, were sufficiently small and GOF (Goodness of fit) was sufficiently close to 1 [10] indicating the refined structural parameters were very reliable.

Figure 2 shows the iron occupancy changes at 8*a*, 16*c* and 16*d* sites as the increase of relaxation time after termination of lithium insertion. The iron occupancy of 8*b* site was almost zero and is not shown in Fig. 2. At lithium insertion process, iron occupancy of 8*a* site decreased and that of 16*c* site increased. After the lithium insertion, the iron occupancy of 8*a* site increased and that of 16*c* site decreased gradually with relaxation time. The iron occupancy of 16*d* site was almost constant all over the process. It was indicated that iron moved from 8*a* site to 16*c* site in the process of lithium insertion and that moved from 16*c* site to 8*a* site at the relaxation process after the lithium insertion. It was considered that lithium was inserted at 8*a* site and iron migrated from 8*a* site to 16*c* site at the lithium insertion process and that, after the lithium insertion, lithium migrated from 8*a* site to 16*c* site and iron moved from 16*c* site to 8*a* site. These are consistent to our previous results that lithium prefers 8*a* site to occupy kinetically and prefers 16*c*

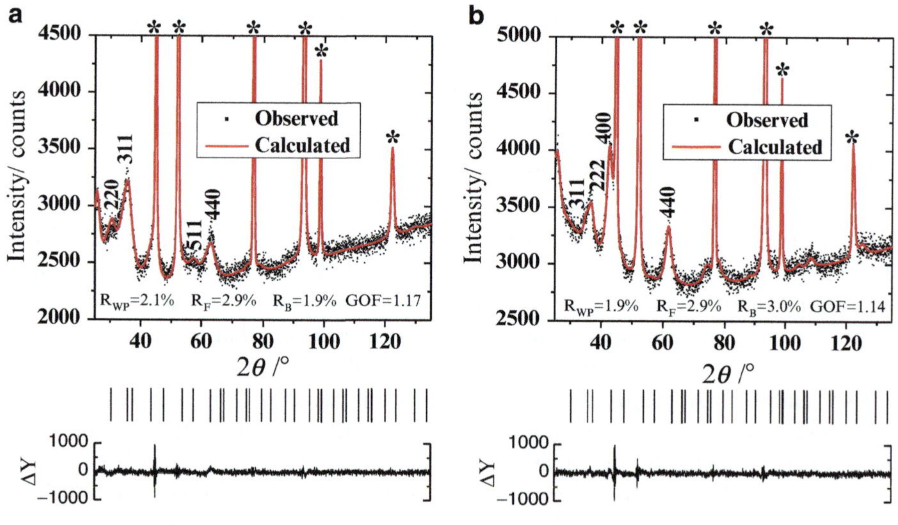

Fig. 1 Rietveld results. (**a**) γ-Fe_2O_3 before lithium insertion. (**b**) Relaxation time of 5 h after lithium insertion (0.01 Ag^{-1} sample for lithium insertion). Calculated and observed patterns are shown in the *top* part by the *solid line* and the *dots*, respectively. The *vertical marks* in the *middle* part show positions calculated for Bragg reflection. The trace in the *bottom* part is a plot of the difference between calculated and observed patterns. *Asterisk* mark indicates the peaks for Ni mesh collector

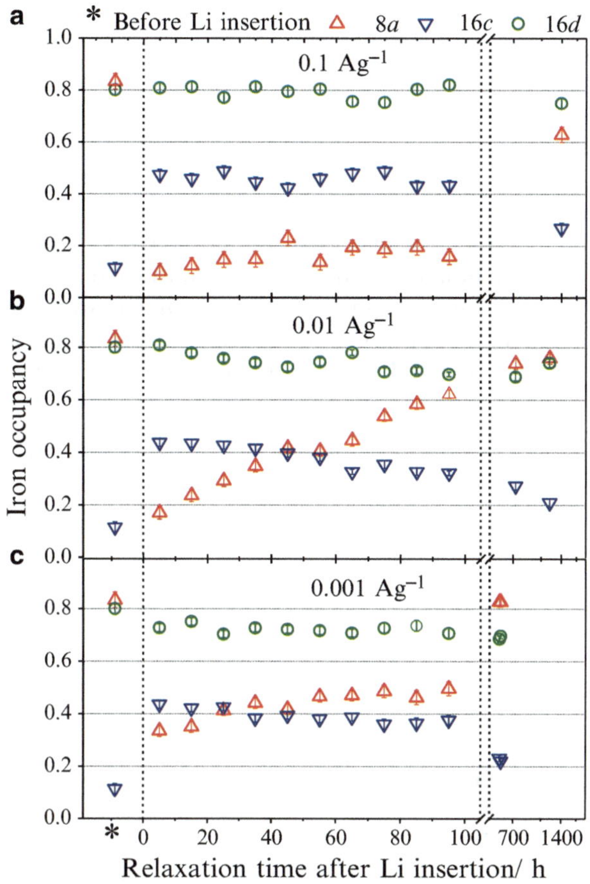

Fig. 2 Iron occupancy changes at 8*a*, 16*c* and 16*d* sites as the increase of relaxation time after termination of lithium insertion: (**a**) 0.1 Ag^{-1}, (**b**) 0.01 Ag^{-1}, (**c**) 0.001 Ag^{-1}. The iron occupancy of 8*b* site was almost zero and is not shown. *Asterisk* mark indicates the sample before lithium insertion and *bars* indicate the estimated standard deviations

site thermodynamically [1, 2]. Lithium is surrounded by four oxygen at 8*a* site and by six oxygen at 16*c* site. It is considered that the smaller number of coordination will make the diffusion easier and the larger number of coordination will make the energy lower. Indeed, it was demonstrated that the lattice energy for lithium distribution at 16*c* site showed lower value than that for lithium distribution at 8*a* site by about 1.1×10^1 kJ mol^{-1} from our first principle calculation [2].

As the decrease of lithium insertion rate, increase of iron occupancy at 8*a* site was observed just after the lithium insertion. It was considered that, for low insertion rate, insertion time is long enough for lithium to migrate from 8*a* site to 16*c* site even during lithium insertion process. These results were well consistent with above consideration that lithium prefers 8*a* site to occupy kinetically and prefers 16*c* site thermodynamically.

4 Conclusion

By the lithium insertion, iron migrated from 8*a* site to 16*c* site and, after the lithium insertion, iron moved from 16*c* site to 8*a* site at the relaxation time. It was considered that lithium was inserted at 8*a* site and iron migrated from 8*a* site to 16*c* site at the lithium insertion process and that, after the lithium insertion, lithium migrated from 8*a* site to 16*c* site and iron moved from 16*c* site to 8*a* site. As the decrease of lithium insertion rate, increase of iron occupancy at 8*a* site was observed just after the lithium insertion. It was considered that, for low insertion rate, discharge time is long enough for lithium to migrate from 8*a* site to 16*c* site even during lithium insertion process. Site preference that lithium prefers 8*a* site to occupy kinetically and prefers 16*c* site thermodynamically was confirmed.

Acknowledgment This work was partly supported by an "Energy Science in the Age of Global Warming" of Global Center of Excellence (G-COE) program (J-051) of the Ministry of Education, Culture, Sports, Science and Technology of Japan and partly supported by Grant-in-Aid for Challenging Exploratory Research (24656581) of the Ministry of Education, Culture, Sports, Science and Technology (MEXT).

References

1. Park S, Oda M, Yao T (2011) Relaxation structure analysis of Li inserted γ-Fe_2O_3. Solid State Ionics 203:29–32
2. Park S, Ito S, Takasu K, Yao T (2012) Multistage Li insertion and extraction relaxation analysis of γ-Fe_2O_3. Electrochem 80(10):804–807
3. Park S, Kameyama K, Yao T (2012) Relaxation crystal analysis of $LiFePO_4$ cathode for Li-ion secondary battery. Electrochem Solid State Lett 15–4:A49–A52
4. Seo IS, Park S, Yao T (2012) Relaxation phase analysis of $LiMn_2O_4$ cathode for secondary Li ion battery. In: Yao T (ed) Zero-carbon energy Kyoto 2011, Green energy and technology. Springer, Japan, pp 165–170
5. Kanamura K, Naito H, Yao T, Takehara Z (1996) Structure change of the $LiMn_2O_4$ spinel structure induced by extraction of lithium. J Mater Chem 6(1):33–36
6. Hibino M, Terashima J, Yao T (2007) Reversible and rapid discharge–charge performance of γ-fe_2o_3 prepared by aquous solution method as the cathode for lithium-ion battery. J Electrochem Soc 154:A1107–A1111
7. Yamauchi S, Hibino M, Yao T (2011) Structure change analysis in γ-fe_2o_3/carbon composite in the process of electrochemical lithium insertion. Solid State Ionics 191:45–48
8. Yao T, Ito T, Kokubo T (1995) Effect of Mn valence on crystal structure of La-Mn-O perovskite oxides. J Mater Res 10:1079–1082
9. Yao T, Ozawa N, Aikawa T, Yoshinaga S (2004) Analysis of layered structures of lithium–graphite intercalation compounds by one-dimensional Rietveld method. Solid State Ionics 175:199–202
10. Young RA (1995) In: Young RA (ed) The Rietveld method. Oxford University Press, Oxford, pp 21–24

Thermal Stability and Catalytic Performance of Pd, Pt and Rh Loaded on $CuO–CeO_2–Al_2O_3$ Support for Three-Way Catalysts

Nguyen The Luong, Eiji Yamasue, Hideki Okumura, and Keiichi N. Ishihara

Abstract The thermal stability and catalytic performance of Pd, Pt and Rh loaded on $CuO–CeO_2–Al_2O_3$ (two layers structure) were investigated to compare with those noble metals coated on $CuO–CeO_2/Al_2O_3$/substrate (three layers structure). CO, C_3H_8 and NO conversion ratios of two layers structure were better than those conversion ratios of three layers structure. The two layers structure was also investigated by means of X-ray diffractometry (XRD), scanning electron microscopy (SEM), energy dispersive X-ray (EDX) and Brunauer-Emmett-Teller (BET). It was clearly shown that, Al_2O_3 introduced into $CuO–CeO_2$ behaves as wall material to prevent the migration of $CuO–CeO_2$.

Keywords Catalytic performance • $CuO–CeO_2–Al_2O_3$ • Noble metals • Thermal stability • Three-way catalysts (TWCs)

1 Introduction

The three way catalysts (TWCs) are capable of simultaneously converting CO, hydrocarbon (HC) and nitrogen oxides (NOx) into harmless CO_2, H_2O and N_2. In TWCs noble metals, such as platinum, rhodium and palladium act as the active components. Oxygen storage capacity (OSC) is one of the crucial factors for the performance of TWCs, and the higher OSC promotes the better dynamic performance of catalysts in converting CO, HC and NOx under conditions from rich to lean oxygen of air-fuel ratios (A/F) in automobile. $CeO_2–ZrO_2$ solid solution is well-known as an excellent supporter for OSC [1]. CeO_2 exhibits oxygen storage/release behavior by the redox reaction of Ce ions between Ce^{3+} and Ce^{4+} [2], and the

N.T. Luong (✉) • E. Yamasue • H. Okumura • K.N. Ishihara
Department of Socio-Environmental Energy Science, Graduate School of Energy Science, Kyoto University, Yoshida Honmachi, Sakyo-ku, Kyoto 606-8501, Japan
e-mail: luong@social-system.energy.kyoto-u.ac.jp

T. Yao (ed.), *Zero-Carbon Energy Kyoto 2012*, Green Energy and Technology, DOI 10.1007/978-4-431-54264-3_19,

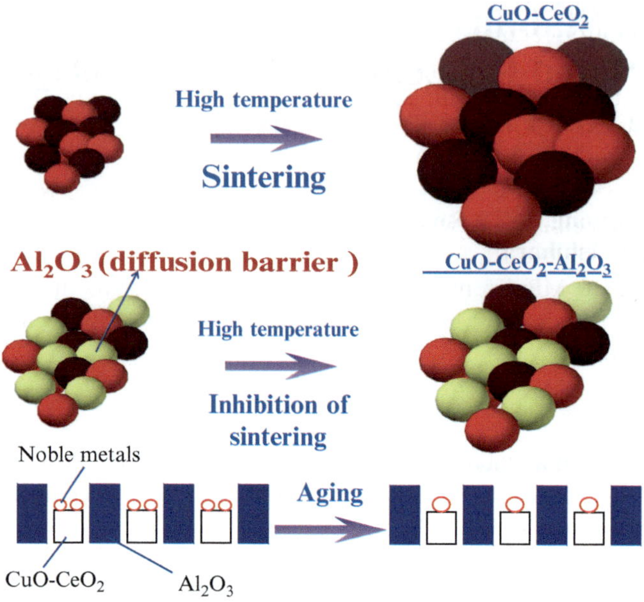

Fig. 1 Illustration of "diffusion barrier concept"

introduction of ZrO_2 into CeO_2 improves the reduction temperature of CeO_2 through structural modification of CeO_2 [3]. Among many studies on CeO_2 base materials such as CeO_2–Al_2O_3 [4], CeO_2–SiO_2 [5], CeO_2–La_2O_3 [4, 6, 7], CeO_2–TbO_x [8], and CeO_2–PrO_x [9] have been reported to improve OSC and increase the thermal stability.

Considering upcoming strict regulation, the TWCs were close-coupled catalyst (CCCs) to engine as a solution to improve the efficiency, so that the temperature in a close-coupled catalyst converter can rise even above 1,000°C. In order to adapt to high temperature, many reports based on performance and durability of CeO_2–ZrO_2 were shown [10–13]. Morikawa et al. [10] reported a new concept of introducing alumina into ceria-zirconia solid solution (ACZ) which was name the "diffusion barrier concept" to enhance OSC. Nakamura [13] developed new concept to prevent migration of noble metal particles and thereby promote the catalytic performance.

Recently, a ceria-copper oxide compound (CuO–CeO_2) prepared by means of high energy mechanical milling has been found to promote oxygen storage capacity (OSC) by authors [14]. The catalytic performance of Pd, Pt and Rh coated on CuO–CeO_2/Al_2O_3/FeCrAl substrate showed improvable CO conversion at lower temperatures and the improvement is significant for CO and C_3H_8 conversion ratios in case lean oxygen [15], from that CuO–CeO_2 can be recommended to apply for TWCs.

It is also extremely important to improve a durability of CuO–CeO_2 to adapt to high temperature. The aim of this study is to develop "diffusion barrier concept" for CuO–CeO_2 structure (Fig. 1), which means that Al_2O_3 introduced into CuO–CeO_2

behaves as a diffusion barrier to prevent migration of $CuO–CeO_2$, thereby suppressing $CuO–CeO_2$ particle growth (Fig. 1). The thermal stability and catalytic performance of Pd, Pt and Rh loaded on $CuO–CeO_2–Al_2O_3$/FeCrAl substrate are investigated and compared with those noble metals loaded on $CuO–CeO_2/Al_2O_3$/ FeCrAl substrate after sintering at 1,000°C for 20 h.

2 Experiments

2.1 Catalysts Preparation

γ-Al_2O_3 (Nilaco Corporation), CuO and CeO_2 (Kojundo Chemical) are used as starting materials. The mixed CuO and CeO_2 powder (50 mol% of CuO content) was milled using a high-energy vibratory ball milling up to 18 h (dry milling). No phase change was observed for CeO_2, while CuO was partially reduced to Cu [4]. For request to create good adhesion, the catalyst slurry contained ~60 wt.% γ-Al_2O_3 and 40% milled $CuO–CeO_2$ powders were further milled with water and a small amount of $Al(NO_3)_3$ as the binder for 24 h (wet milling). The catalyst slurry was subsequently dip-coated on Al_2O_3 layer thin film thermally precipitated on the FeCrAl substrate (the dimension 3 × 18 mm), followed by sintering for 2 h in air at 250°C and then for 2.5 h at 650°C, the details of deposition of mixed $CuO–CeO_2–\gamma$-Al_2O_3 washcoat on FeCrAl substrate were referred in [16]. The porous $CuO–CeO_2–\gamma$-Al_2O_3 washcoat was impregnated with the mixed solution of Pt $(NO_3)_2$, $Pd(NO_3)_2$ and $Rh(NO_3)_3$ to be the ratio of Pt:Pd:Rh = 1:14:1 (total 3.7 g/L) [1]. The total loading amount of noble metal was 4 wt.% of CeO_2–CuO and kept the same weight level for the all samples.

2.2 Surface Characterization

The structure and morphology of catalyst layer were analyzed by X-ray diffractometry (XRD) using Cu–Kα radiation (RIGAKU RINT-2100CMT), scanning electron microscope (SEM, JEOL JSM-5800) and by energy dispersive X-ray (EDX). The surface area was estimated by N_2 adsorption method (Brunauer-emmett-Teller).

2.3 Catalytic Activity Measurements (CO, C_3H_8, NO)

The simulant exhaust gas containing O_2, CO (1.5%), H_2 (0.5%), CO_2 (12%), C_3H_8 (0.1%), NO (0.05%) and N_2 (balance) was prepared, the λ value as oxidants/ reductants factor was defined to be $\lambda = (2O_2 + NO)/(CO + H_2 + 10C_3H_8)$.

The λ was adjusted by controlling the concentration of oxygen, catalyst activity (CO, C_3H_8, NO) was analyzed by GC-TCD (Varian CP-4900), the sample was put in a reaction tube (i.d = 8 mm) made from quartz. In case $\lambda = 1$ (the concentration of O_2 was 1.5%), the reaction temperature was varied from 30°C to 540°C at the heating rate of 3°C/min with the gas flow of 20 mL/min, the catalytic performance was measured with the increase of temperature. In case $\lambda < 1$, the concentration of oxygen was controlled from 1.5% ($\lambda = 1$) to 0.42% ($\lambda = 0.3$) by oxygen observer agent (227 g of ageless) and absorber time about for 15 min, the reaction temperature was kept constant at 500°C, the catalytic activities were measured with the $\lambda = 0.85$.

3 Results

3.1 The Structure Characterization of Catalysts

XRD patterns of CuO–CeO_2–γ-Al_2O_3 layer coated on FeCrAl substrate was shown in Fig. 2a, b respectively. The peaks of Fe–Cr–Ni substrate, γ-Al_2O_3, CuO and CeO_2 were observed after sintering at 650°C for 2.5 h (Fig. 2a), the intensity peaks of Fe–Cr–Al, γ-Al_2O_3, CuO and CeO_2 became sharper after aging at 1,000°C for

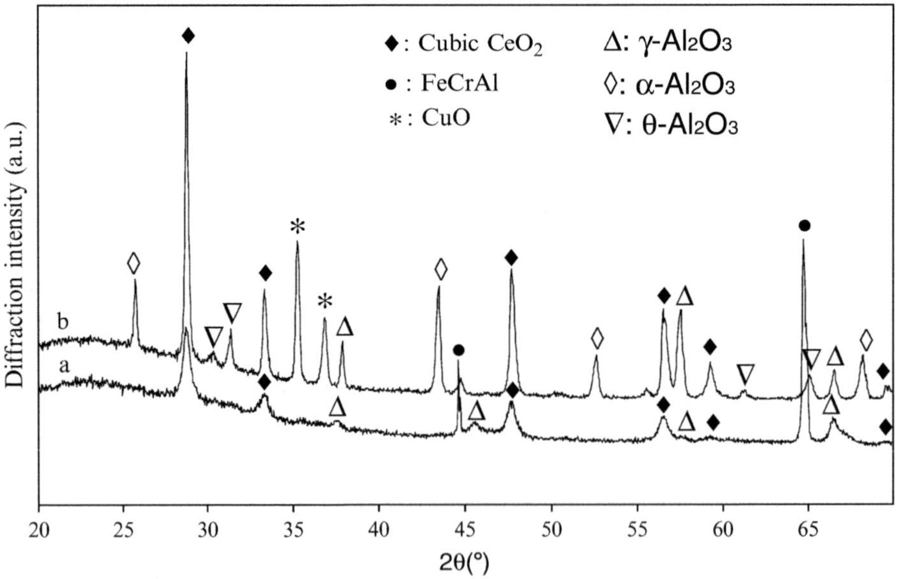

Fig. 2 XRD patterns of CuO–CeO_2–Al_2O_3 coated on FeCrAl substrate calcined at (**a**) 650°C and (**b**) 1000°C for 20 h

Table 1 The surface area of two and three layers structure after sintering at 650°C for 2.5 h and at 1,000°C for 20 h

No	Samples	BET (cm^2/cm^2 substrate)
1	$CuO–CeO_2/\gamma-Al_2O_3$/substrate-650°C for 2.5 h	14–16
2	$CuO–CeO_2/\gamma-Al_2O_3$/substrate-1,000°C for 20 h	5–7
3	$CuO–CeO_2–\gamma-Al_2O_3$/substrate-650°C for 2.5 h	15–17
4	$CuO–CeO_2–\gamma-Al_2O_3$ substrate-1,000°C for 20 h	6–8

Fig. 3 SEM photographs of $CuO–CeO_2–\gamma-Al_2O_3$ layer on FeCrAl substrate after sintering at 650°C for 2.5 h

20 h, no phase changes of CuO and CeO_2 were observed, while the new phases of $\alpha-Al_2O_3$ and $\theta-Al_2O_3$ were observed after sintering at 1,000°C for 20 h (Fig. 2b).

Table 1 shows that, the surface area of the mixed $CuO–CeO_2–Al_2O_3$ coated on substrate and $CuO–CeO_2$ coated on $\gamma-Al_2O_3$ washcoat on substrate decreased after sintering at 1,000°C for 20 h. The reason for this, the phase changes of $\gamma-Al_2O_3$ to $\alpha-Al_2O_3$ and $\theta-Al_2O_3$ were observed after sintering at 1,000°C for 20 h. The surface area of mixed $CuO–CeO_2–Al_2O_3$ coated on substrate was similar with $CuO–CeO_2$ coated on $\gamma-Al_2O_3$ washcoat on substrate after sintering at 650°C for 2.5 h and at 1,000°C for 20 h.

The morphology of $CuO–CeO_2–Al_2O_3$ layer is shown in Fig. 3. As the results of milling process of $CuO–CeO_2–\gamma-Al_2O_3$ slurry, the fine agglomerate particles of $CuO–CeO_2$ after sintering at 650°C were observed. The distribution of fine $CuO–CeO_2$ particles can be seen in the dense microstructure of Al_2O_3 with dimensions less than 2 μm. Energy dispersive X-ray (EDX) analysis with color mapping of the $CuO–CeO_2–Al_3O_3$ coating samples is shown in Fig. 4 which reveals that the contents are limited to Ce, Cu, Al and oxygen. The contaminations from the milled media (ZrO_2 balls and a steel vial) are also examined by EDX, no contamination like Fe, Cr or ZrO_2 was detected while C and Au were observed as pre-treatment step of SEM. The color mapping of EDX analysis shows homogeneous distributions of CuO, CeO_2 and Al_2O_3.

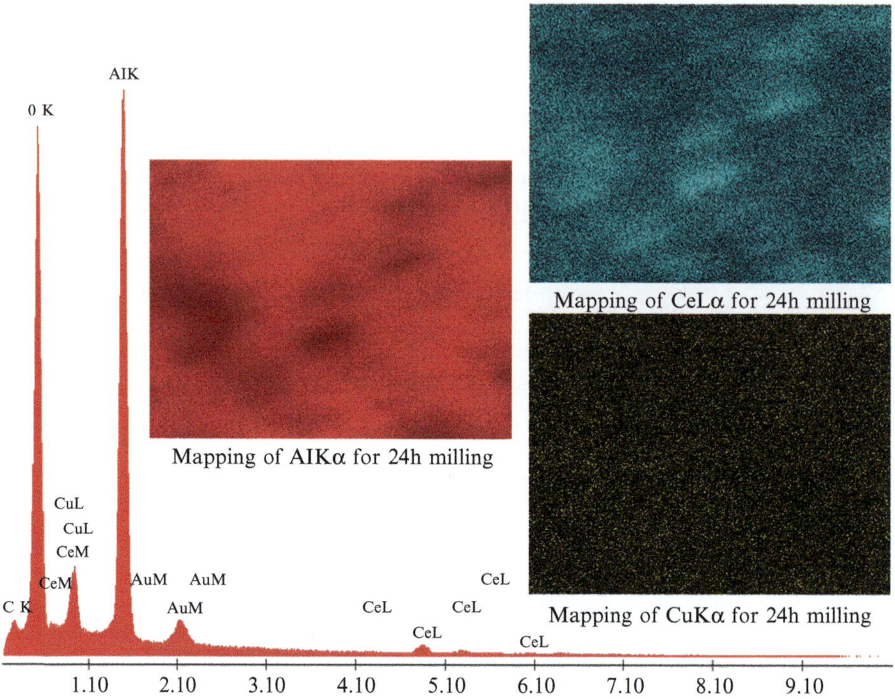

Fig. 4 Energy dispersive X-ray spectrum of the $CuO–CeO_2–Al_2O_3$ coating layer after sintering at 650°C, the $CuO–CeO_2–Al_2O_3$ slurry was milled for 24 h, showing the distribution of Cu, Ce, Al on the coating layer

3.2 Catalytic Performance

3.2.1 Catalytic Performance in Case $\lambda = 1$

Figure 5 shows the catalytic performance of noble metals coated on $CuO–CeO_2–Al_2O_3$/substrate after sintering at 650°C for 2.5 h and at 1,000°C for 20 h. In case sintering at 650°C for 2.5 h, the CO, C_3H_8 and NO conversion ratios was increased from 0% to 99%, 62.1% and 84.7%, respectively at temperature range from 30°C to 540°C. The results showed that (not shown here), the similar CO, C_3H_8 and NO conversion ratios were observed for noble metals coated on $CuO–CeO_2/Al_2O_3$/substrate after sintering at 650°C for 2.5 h.

Figure 6 shows CO, C_3H_8 and NO conversion ratios at 500°C, under enough oxygen condition ($\lambda = 1$) and over catalysts of noble metals coated on $CuO–CeO_2–Al_2O_3$/substrate and those noble metals coated on $CuO–CeO_2/Al_2O_3$/substrate after sintering at 1,000°C for 20 h. The results showed CO, C_3H_8 and NO

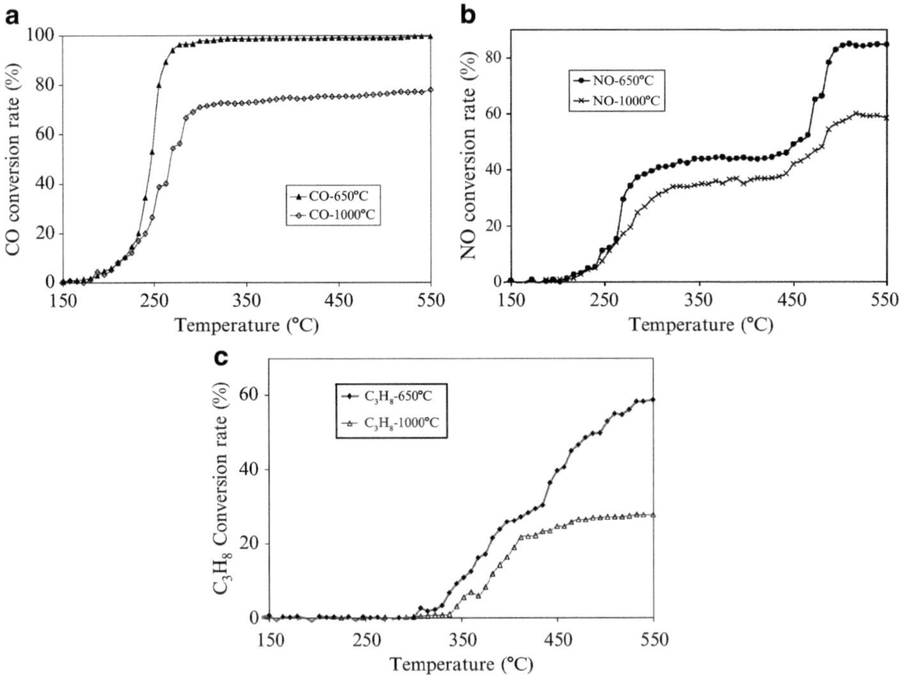

Fig. 5 (**a**) CO; (**b**) C_3H_8, (**c**) NO conversion performance of noble metals coated on $CuO–CeO_2–Al_2O_3$/substrate, after sintering at 650°C for 2.5 h and sintering at 1000°C for 20 h

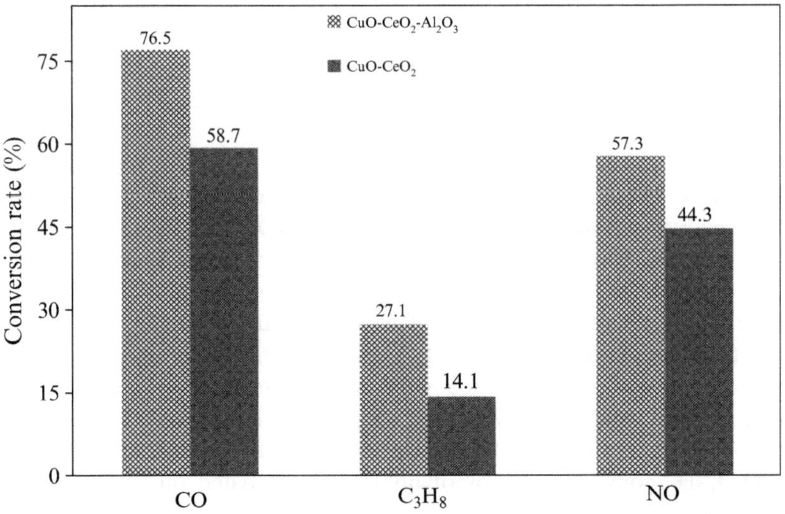

Fig. 6 CO, C_3H_8 and NO conversion performance at 500°C of noble metals coated on $CuO–CeO_2–Al_2O_3$/substrate and those noble metals coated on $CuO–CeO_2/Al_2O_3$/substrate after sintering at 1000°C for 20 h, $\lambda = 1$

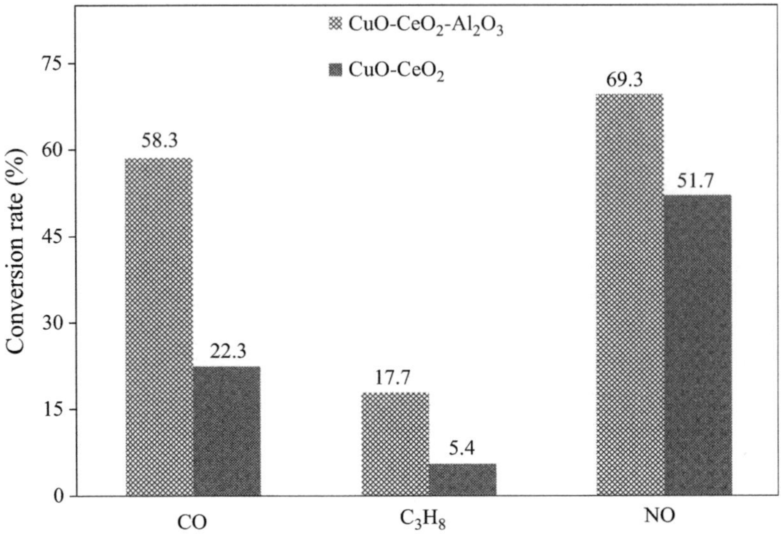

Fig. 7 CO, C_3H_8 and NO conversion performance at 500°C of noble metals coated on CuO–CeO_2–Al_2O_3/substrate and those noble metals coated on CuO–CeO_2/Al_2O_3/substrate after sintering 1000°C for 20 h, in case learn oxygen ($\lambda = 0.85$)

conversion ratios of noble metals coated on CuO–CeO_2–Al_2O_3/substrate were higher than those conversion ratios of noble metals coated on CuO–CeO_2/Al_2O_3/substrate after sintering at 1,000°C for 20 h.

3.2.2 Catalytic Performance in Case Lean Oxygen ($\lambda < 1$)

Figure 7 shows CO, C_3H_8 and NO conversion ratios at 500°C, under lean oxygen condition ($\lambda < 1$) and over catalysts of noble metals coated on CuO–CeO_2–Al_2O_3/substrate and those noble metals coated on CuO–CeO_2/Al_2O_3/substrate after sintering at 1,000°C for 20 h. Compare catalytic performance with in case $\lambda = 1$ and $\lambda < 1$, the results showed CO and C_3H_8 conversion ratios of both two layers and three layers structure in case $\lambda < 1$ were lower than CO and C_3H_8 conversion ratios of those structure in case $\lambda = 1$, while NO conversion ratios was higher.

Compare catalytic performance with two layers structure and three layers structure at 500°C and in case $\lambda < 1$ (0.85), the results showed CO, C_3H_8 conversion ratios of noble metals coated on CuO–CeO_2–Al_2O_3/substrate is much higher than CO, C_3H_8 conversion ratios of noble metals coated on CuO–CeO_2/Al_2O_3/substrate (Fig. 7).

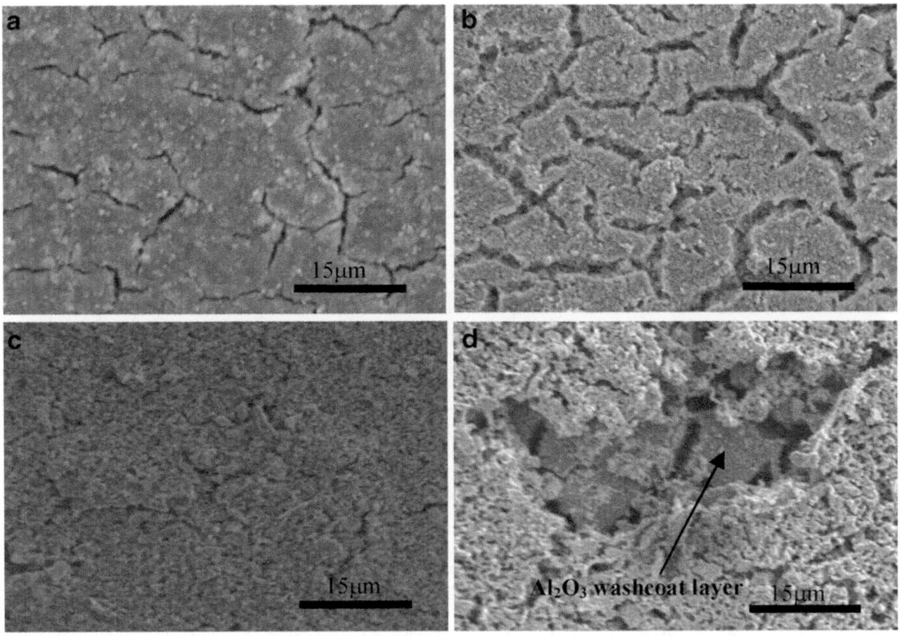

Fig. 8 Morphology of (**a**) the CuO–CeO_2–Al_2O_3 layer coated on FeCrAl substrate after sintering at 650°C for 2.5 h and (**b**) calcined at 1000°C for 20 h; (**c**) the CuO–CeO_2 coated on Al_2O_3 on FeCrAl substrate sintered at 650°C for 2.5 h and (**d**) calcined at 1000°C for 20 h

4 Discussion

In this section, the role of diffusion barrier on structure of CuO–CeO_2–Al_2O_3/substrate will be discussed.

In the order to show the effect of diffusion barrier, the morphology of CuO–CeO_2–Al_2O_3 on substrate (two layers) was used to compare with the morphology of CuO–CeO_2 on Al_2O_3 washcoat layer on substrate (three layers). Figure 8a–c show the similar particle size of CuO–CeO_2 for both two and three layers structure after sintering at 650°C for 2.5 h. After sintering at 1,000°C for 20 h, the particle size of CuO–CeO_2 became larger. In case of CuO–CeO_2–Al_2O_3 coated on substrate, Al_2O_3 introduced into CuO–CeO_2 behaves as a diffusion barrier to prevent migration of CuO–CeO_2, thereby suppressing CuO–CeO_2 particle growth. The smaller CuO–CeO_2 particles were observed (Fig. 8b). In case of CuO–CeO_2 coated on Al_2O_3, CuO–CeO_2 particles were easy to aggregate, thereby the accelerating CuO–CeO_2 particles was larger and appearing space of Al_2O_3 washcoat layer on the surface were observed (Fig. 8d). It is found that the particle size of CuO–CeO_2 is significantly reduced after sintering at 1,000°C for 20 h for two layers structure.

It also reported that [13], CeO_2 was strong chemical bonding with noble metals which was used as material support, while γ-Al_2O_3 was weak chemical bonding with noble metals which was used as wall material to inhibit aggregation of material support particles (CuO–CeO_2), since the precious metal particles do not migrate over the wall material. As a result, the precious metal particles were kept in the smaller particles size, thereby the catalytic performance of two layers structure was higher than catalytic performance of three layers structure after sintering at 1,000°C for 20 h (Fig. 6). In case learn oxygen, the catalytic performance depend strongly on OSC. In two layer structure, Al_2O_3 introduced into CuO–CeO_2 behaves as a diffusion barrier to prevent migration of CuO–CeO_2, thereby suppressing CuO–CeO_2 particle growth. The OSC of CuO–CeO_2–Al_2O_3/substrate improved after sintering at 1,000°C for 20 h, therefore, the CO, C_3H_8, conversion ratios of two layers structure were higher than those conversion ratios of three layers structure after sintering at 1,000°C for 20 h (Fig. 7).

5 Conclusion

The catalytic performance of CuO–CeO_2–Al_2O_3 loaded on Pd, Pt and Rh was investigated after sintering at 650°C for 2.5 h and 1,000°C for 20 h.

In CuO–CeO_2–Al_2O_3 structure, Al_2O_3 introduced into CuO–CeO_2 behaves as wall material to prevent migration of CuO–CeO_2 during sintering at 1,000°C for 20 h.

As results of inhibition, the structure of CuO–CeO_2–Al_2O_3 is more stable than CuO–CeO_2 layer coated on Al_2O_3 at high temperature. CuO–CeO_2 particle sizes and the precious metal particle size are kept smaller, so that the CO, C_3H_8, NO conversion ratios of two layers structure was higher than those conversion ratios of three layers structure after sintering at 1,000°C for 20 h.

Acknowledgement This work was supported by Global Center of Excellence (GCOE) Program and the Monbukagakusho Scholarship, Japan.

References

1. Kaspar J, Fornasero P, Hickey N (2003) Automotive catalytic converters: current status and some perspectives. Catal Today 77:419–449
2. Ivanova AS (2009) Physicochemical and catalytic properties of systems based on CeO_2. Kinet Catal 50(6):797–815
3. Gamarra D, Munuera G, Hungria AB, Fernandez-Garcia M, Conesa JC, Midgley PA, Wang XQ, Hanson JC, Rodriguez JA, Martinez-Arias (2007) Structure–activity relationship in nanostructured copper–ceria-based preferential co oxidation catalysts. J Phys Chem C 111:11026–11038
4. Usmen RK, Graham GW, Watkins WLH, McCabe RW (1995) Incorporation of La^{3+} into a Pt/CeO_2/Al_20_3 catalyst. Catal Lett 30:53–63

5. Bensalem A, Bozon-Verduraz F, Delamar M, Bugli G (1995) Preparation and characterization of highly dispersed silica-supported ceria. Appl Catal A 121:81–93
6. Miki T, Ogawa T, Haneda M, Kakuta N, Ueno A (1990) Enhanced oxygen storage capacity of cerium oxides in CeO2/La2O3/Al2O3 containing precious metals. J Phys Chem 94:6464–6467
7. Colón G, Navío JA, Monaci R, Ferino I (2000) CeO2–La2O3 catalytic system Part I. Preparation and characterisation of catalysts. Phys Chem Chem Phys 2:4453–4459
8. Yei F, Mori T, Ou DR, Zou J, Drennan J (2008) Microstructural characterization of $Ce_{1-x}Tb_xO_{2\_\delta}$ ($0.60 \leq x \leq 0.90$) sintered samples. Mater Res Bull 43:759–764
9. Sinev MY, Graham GW, Haack LP, Shelef M (1996) Kinetic and structural studies of oxygen availability of the mixed oxides $Pr_{1-x}M_xO_y$ (M = Ce, Zr). J Mater Res 11:1960–1971
10. Kanazawa T, Suzuki J, Takada T, Suzuki T (2003) Development of three-way catalyst using composite alumina-ceria-zirconia. SAE Technical Paper 2003-01-0811
11. Morikawa A, Suzuki T, Kanazawa T, Kikuta K, Suda A, Shinio H (2008) A new concept in high performance ceria-zirconia oxygen storage capacity material with Al_2O_3 as a diffusion barrier. Appl Catal B 78:210–221
12. Morikawa A, Kikuta K, Suda A, Shinjo H (2009) Enhancement of oxygen storage capacity by reductive treatment of Al_2O_3 and CeO_2–ZrO_2 solid solution. Appl Catal B 88:542–549
13. Nakamura M, Wakamatsu H, Suga K, Sekiba T, Hiramoto Y, Shibata K (2009) Ultrafine precious catalyst for high conversion efficiency with low precious metal loading. SAE paper No 2009-01-1069
14. Luong NT, Yamasue E, Okumura H, Ishihara KN (2012) Structure and catalytic behavior of CuO-CeO_2 prepared by high-energy ball milling. Catal Lett (under review)
15. Luong NT, Yamasue E, Okumura H, Ishihara KN (2012) Investigation on catalytic performance of Pt, Pd and Rh/CuO-CeO_2/Al_2O_3/metallic substrate support for three-way catalysts (to be appeared)
16. Luong NT, Yamasue E, Okumura H, Ishihara KN (2012) Adhesion properties of milled CuO-CeO_2/γ-Al_2O_3 on metallic substrate for automotive catalytic converter. Particulate Sci Technol (under review)

Part III
Advanced Nuclear Energy Research

(i)
Contributed Paper

Inherently-Safe High Temperature Gas-Cooled Reactor

Masuro Ogawa

Abstract High temperature gas-cooled reactor, also known as VHTR, is one of the six Generation-IV nuclear energy systems. Proven capable of coolant outlet at 950°C, it can generate electricity at 50% thermal efficiency. It can also produce hydrogen and supply high-quality process heat. Its production remains highly-economical even at as small as 600 MW plant size. The greatest potential of the reactor, however, is its nuclear safety. In this report, a new concept of Inherently-safe HTGR is proposed as a countermeasure against accident similar to that occurred on March 11, 2011 at the Fukushima Dai-ichi Nuclear Power Station (F1). The underlying objective of the proposal is that the Inherently-safe HTGR never compromises the safety of the public, society and environment. It is designed so that any accident such as the F1 accidents will be self-controllable to safe state without reliance on active and/or passive devices and systems. What is revolutionary about the Inherently-safe HTGR is that physical phenomena are always present at work in the event of accident to correct any conditions that might result in failure of cladding of uranium fuel. Some inherent-safety characteristics of the Inherently-safe HTGR have already been confirmed by test of the HTTR, currently the world largest HTGR test reactor of 30 MW thermal power and 950°C, developed and constructed by JAEA in Oarai. The balance of the inherent safety characteristics required to establish the Inherently-safe HTGR concept is analyzed. These results of the Inherently-safe HTGR are presented, here.

Keywords Counter physical phenomena • High economy • HTTR • Inherently-safe high temperature gas-cooled reactor (HTGR) • Self controlled characteristics

---

M. Ogawa (✉)
Nuclear Hydrogen and Heat Application Research Center, Japan Atomic Energy Agency,
Naritacho 4002, Oarai-machi, Higashiibaraki-gun, Ibaraki-ken 311-1393, Japan
e-mail: ogawa.masuro@jaea.go.jp

T. Yao (ed.), *Zero-Carbon Energy Kyoto 2012*, Green Energy and Technology,
DOI 10.1007/978-4-431-54264-3_20,

1 Introduction

High Temperature Gas-cooled Reactor (HTGR) is a thermal neutron, graphite-moderated and helium gas-cooled nuclear reactor. The purpose of HTGR is utilization of high temperature heat of 950°C higher than existent Light Water Reactors (LWRs) of 300°C. By using this high temperature heat, not only electricity but also hydrogen and process heat can be produced with high thermal efficiency because thermal efficiency increases with temperature according to the Carnot cycle. Additionally, low-temperature waste heat of 200–300°C can be also supplied for district heating, desalination, agriculture and aquatic product industry as shown in Fig. 1 [1]. So, total heat utilization ratio can be as high as 70–80% of reactor thermal power. Although HTGR is generally sized less than 600 MW thermal power or 300 MW in electricity, the range of a small-sized reactor, it is economically competitive due to mainly the features of multi-purpose, high efficiency and system simplicity including safety system [2]. Because of the small size and because it can be inherently safe, HTGR can be installed adjacent to small- to mid-sized cities not connected to large power grid, especially, in inland areas because a large amount of cooling water is not necessary to operate the HTGR. HTGR is thus suitable as a decentralized energy source, of course, with CO_2 emission free.

Japan Atomic Energy Agency (JAEA) constructed and has operated the Japanese first HTGR, named High Temperature Engineering Test Reactor (HTTR), of which maximum reactor outlet temperature is 950°C and thermal full power is 30 MW [3, 4] as shown in Fig. 2. A 50-day continuous operation with its reactor

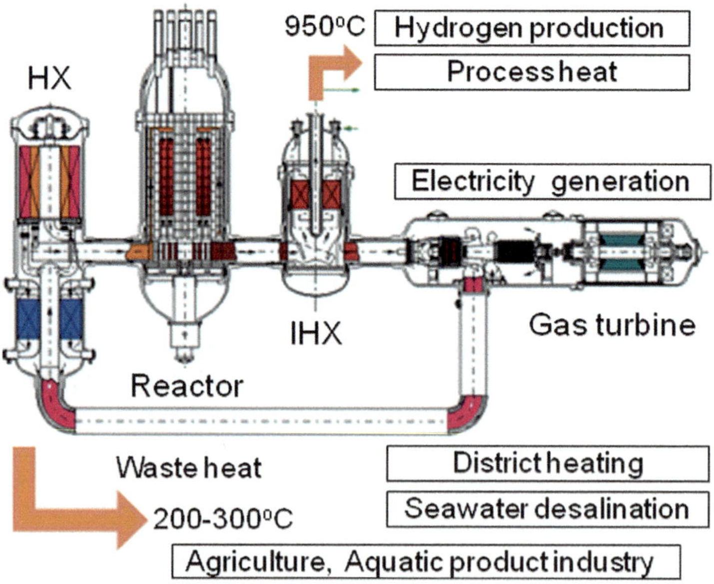

Fig. 1 Various utilization of HTGR

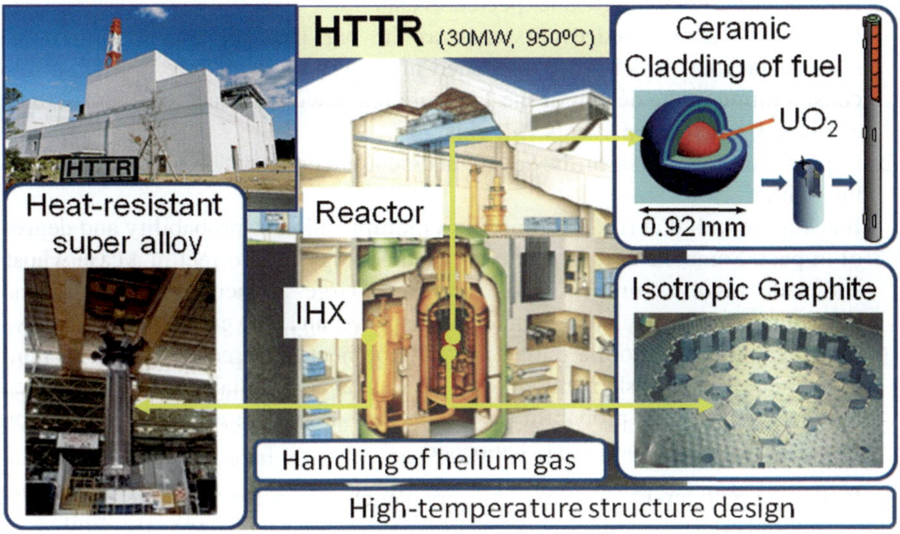

Fig. 2 The Japan's leading edge technologies in HTGR

outlet temperature of 950°C and full thermal power was performed in early 2010 [5]. It is shown that the HTTR is a reliable energy system to provide high temperature heat to non-electric application.

In HTTR, there are mainly five core technologies of HTGR that enable the safe production of high temperature nuclear heat as shown in Fig. 2. Those are ceramic cladding fuel, isotropic graphite, heat-resistant super alloy, high-temperature structure design and handling of helium gas. Cladding of UO_2 fuel is the most important barrier to confine radioactive fission products. This cladding is made of four layers of ceramic coating over a UO_2 fuel kernel. The resulting spherical fuel particle has a diameter of about 1 mm. Graphite is carbon element as well as diamond and amorphous carbon. The difference is that diamond has three dimensions while graphite two dimensions. This dimensional difference affects physical properties such as strength and thermal conductivity. A hot-pressurizing technology manufactures a large graphite block into three dimensional, namely, isotropic structure from two-dimensional one. Hastelloy XR is a fortified nickel-based super alloy of strong heat resistance. Besides use of the super alloy, the high temperature structure design enables the reactor components to operate in high temperature. Because helium, the second lightest element after hydrogen, is leak-prone from pipe and vessel, many experiences have been accumulated in the handling of helium gas.

Many subjects are being pointed out after the accident of Fukushima Dai-ichi Nuclear Power Plants (F1) in Japan on March 11, 2011. One of the subjects for nuclear power plants required to continuously contribute to human being as a power generation plant is how to obtain public reliance on nuclear power plants. In concrete, there are two issues as follows:

1. How do we address "new unexpected events"?
   Even if the unexpected events such as the big Tsunami and the prolonged station-blackout, which have happened in Fukushima and were not taken into consideration in the design of the F1, were redefined to "expected events" to be evaluated, unexpected events still exist.
2. Explanation that "Risk of the severe accident is very low." is not acceptable to the public.
   It is known well that risk is defined as a multiplication of probability and degree of impact. Someone says that the risk methodology is incorrect in a case whose end state involves unrecoverable and very large impact to the public and environment. Namely, you should not approve any power plants which give irretrievable impact to the public and environment. The second reason of non-acceptance of the risk methodology is that an accidental event can happen tomorrow, even if its probability is as very low as one per a million or a billion years. Accentually, the F1 has happened.

In the present report, a safety concept which can solve the above issues is proposed on HTGR.

2 Proposal

Even if any unexpected accidents, so called design extension conditions (DECs; DEC is recently used instead of "beyond design based accident"), happen and the nuclear power plant falls into an uncontrollable state, so as not to extremely affect the public and environment, I propose a nuclear power plant being able to attain inherent safety with only self-control features based on counter physical phenomena, but not with any active and/or passive devices and facilities. The counter physical phenomena happen deterministically but not probabilistically. Thus, counter physical phenomena is not argued with probability.

What are countermeasures on safety of nuclear power plants in a post-crisis Japan?

If man cannot control the system like the F1 accident, first (1): Man can promptly repair or replace the broken systems with active safety devices and facilities.

If man cannot control the system, man can rely on a self-control system, that is, system to control it, second (2): With built-in passive devices and facilities, third (3): Without active and/or passive devices and facilities.

The second issue is a passive way, while the third issue is an inherent way to use only counter physical phenomena such as Doppler effect, thermal radiation and natural convection of the atmosphere but not reactor coolant.

The inherent safety is the supreme goal of nuclear reactor power plants. HTGR could take the way of inherent safety. Figure 3 shows a new concept of Inherently-safe HTGR proposed here. Physical events to lose confinement function of cladding consist of "diffusion" of fission product (FP) passing through cladding at high temperature, "melting or sublimation" of cladding, chemical "corrosion" of cladding, and mechanical "rupture." Causes events of these physical events include

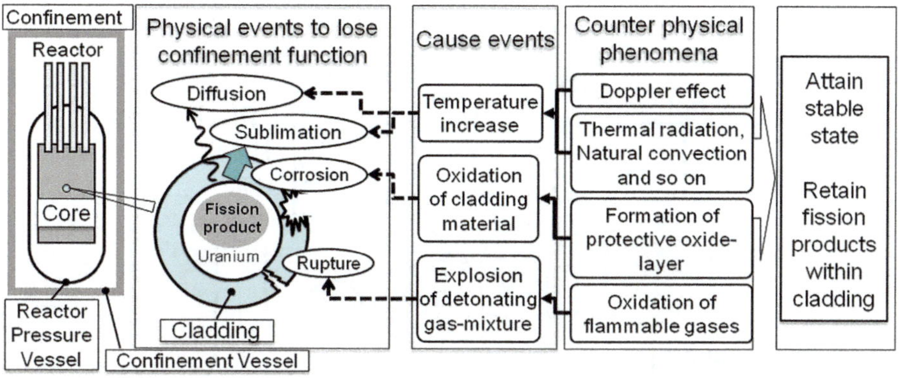

Fig. 3 Proposal of new concept on inherently-safe HTGR

"temperature increase" due to heating-increase or cooling-decrease, "oxidation of cladding material" due to massive air ingress, and "explosion of detonating gas-mixture" such as CO and oxygen due to chemical reaction between graphite and corrosive gases in the HTGR core. If counter physical phenomena are present at work to restrain the cause events, HTGR can attain stable state and retain fission products within cladding, thus, no harmful impact on the public and environment, and possibly no need of public evacuation. These counter physical phenomena are naturally occurring in that they do not fail under any conditions.

This is the essence of the new inherent safety concept. The present concept of Inherently-safe HTGR can respond the two issues described in the first chapter in the following. For the first subject of "new unexpected events," even if any unexpected accidents beyond design based accidents happen and the nuclear power plant falls into an uncontrollable state impossible of reuse, Inherently-safe HTGR does not remarkably affect the public and environment with only self-control features, that is, only counter physical phenomena, but without any active and/or passive devices and facilities. For the second risk subject, counter physical phenomena happen deterministically but not probabilistically. Thus, counter physical phenomena is not argued with probability, as never was a failure of daily sunrise and sunset known since the mankind. HTGR is certainly one of the most promising nuclear reactors having inherently-safe features.

3 Verification of Counter Physical Phenomena in Inherently-Safe HTGR

Can we design a nuclear reactor so as to restrain temperature increase, oxidation of cladding material and explosion of detonating gas-mixture with only counter physical phenomena naturally occurring, that is, with no active and/or passive

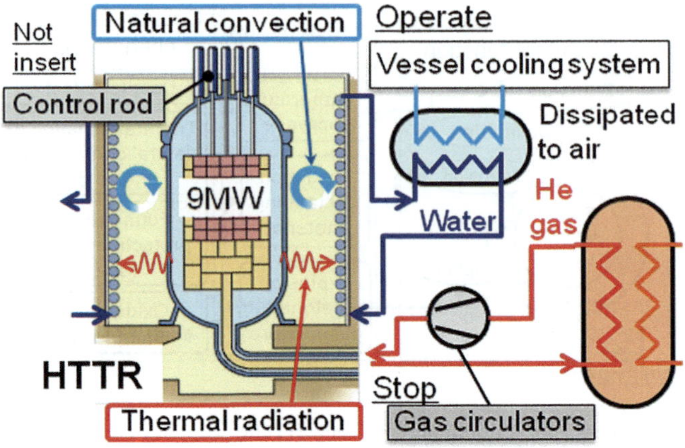

Fig. 4 HTTR safety test (Oct 2010)

devices and facilities? So, it will be analyzed here whether design parameters such as power density, configuration of core and so on can be determined within a feasible range. Temperature increase is caused by increase of heat generation or decrease of heat removal (cooling). In the temperature increase, we have planned verification test with HTTR. And a safety test has been conducted in HTTR on October 2010 for the temperature increase caused by decrease of heat removal as described in Sect. 3.1, and ranges of design parameters will be examined for the cause events as described in Sect. 3.2.

3.1 Experimental Results with HTTR

Before describing the HTTR test, let's review the course of the F1 accident. The F1 automatically scrammed due to the Tohoku earthquake (March 11, 2011), and then nuclear fission and heat generation were stopped. The problem was that the decay heat could not be removed due to the Tsunami, and then the decay heat has just caused multiple-core melt-down, hydrogen explosion and release of radioactive materials to environment in succession. Roughly speaking, decay heat, which is produced by FP, amounts to 7% of full reactor power after 1 s of reactor shutdown, 0.7% after 1 day, 0.07% after 1 year. Even though decay heat is small, failure to remove it from core can increase core temperature high enough to melt down the core.

Even if any cooling systems in HTGR do not work as same as the F1 accident, HTGR can inherently remove the decay heat. This inherent safety characteristics has been demonstrated by test of the HTTR on October 2010, prior to the F1 accident (March 11, 2011) in the following as shown in Fig. 4 [6]. The HTTR test conditions were severer than the F1 accident:

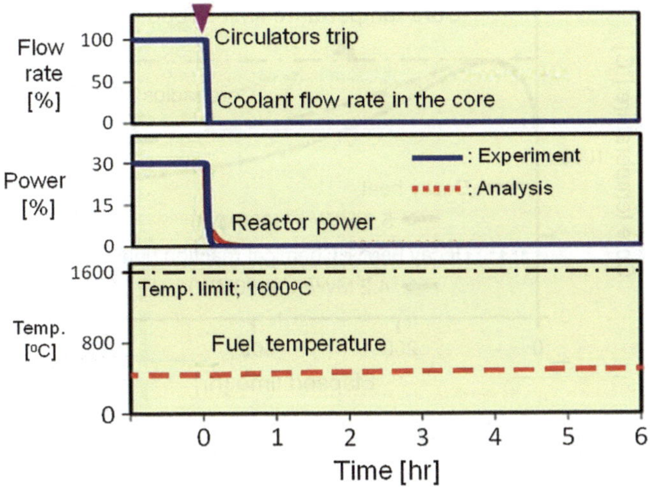

Fig. 5 Experimental and analytical results of HTTR safety test

- Maintain with 30% (9 MW) of the full power (30 MW)
- Stop gas circulators for the primary helium gas coolant
- Simultaneously not insert the control rods into the core
- Continue to operate the vessel cooling system

The experimental and analytical results are shown in Fig. 5. After the stop of gas circulators, the control rods were not inserted. But reactor power nevertheless immediately has decreased down to almost zero. Fuel temperature calculated has remained practically constant and well below the fuel temperature limit of 1,600°C. Doppler effect mainly results in the decrease of reactor power. Doppler effect comes from increasing neutron absorption by U-238 (called resonance capture broadening) with increasing U-238 temperature. Plenty of U-238 is contained inside enriched uranium fuel of 6% in average. It is found that HTTR has been designed so that Doppler effect works well to restrain temperature increase due to loss of flow rate.

After the power decrease, decay heat has been removed by thermal radiation and natural convection of the atmosphere from the outer surface of reactor pressure vessel even without cooling. In this HTTR test, a vessel cooling system surrounding the exterior of the reactor pressure vessel as a final heat sink was operated as shown in Fig. 4 because this safety test was done for the first time in the HTTR. Thus, it is found that the reactor power has decreased down to almost zero after the stop of gas circulators even though control rods were not inserted to the core, and that decay heat was removed from the outer surface of the reactor pressure vessel due to thermal radiation and natural convection of the atmosphere even though with no cooling. Decay heat removal will be described in detail in the next section.

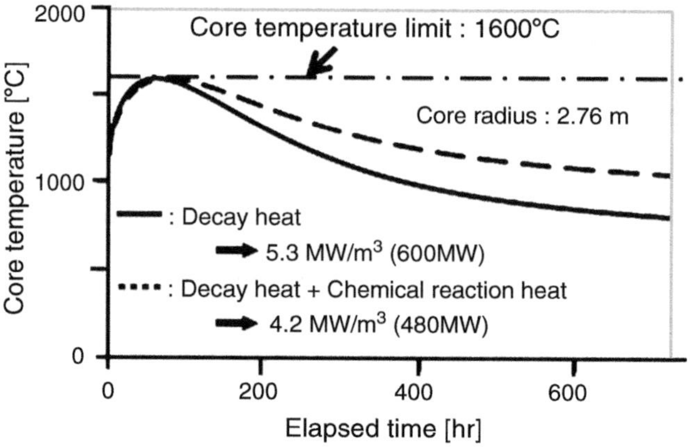

Fig. 6 Change in core temperature after loss of flow rate

Two safety tests are scheduled to be performed from initial 80% and 100% of the full power, and then a HTTR test will be carried out without the running of that vessel cooling system, namely, no cooling systems to simulate whole station blackout. Furthermore, HTTR tests will be planned on the temperature increase due to increase of heat generation which is caused by flow-rate increase or withdrawal of control rods, and on measurement of radioactive materials in the primary coolant system.

3.2 Analytical Results

Ranges of design parameters are examined here, in which counter physical phenomena which work to restrain cause events of temperature increase, oxidation of cladding material and explosion of detonating gas-mixture can attain a stable state of the reactor and can retain FP within the cladding materials even if any accidents would happen.

Figure 6 shows change in core temperature with elapsed time after loss of flow rate of primary helium gas coolant [7]. At the elapsed time = 0 h, reactor power was decreased down to almost zero mainly by Doppler effect, and then core temperature heats up and begins to cool-down as elapsed time. The initial excess heat generation is temporary stored in the massive graphite core structure. As the decay heat decreases, core cool-down begins by thermal radiation and natural convection from the surface of reactor pressure vessel. Under conditions of core radius = 2.76 m and core temperature limit = 1,600°C, maximum power density is calculated to 5.3 MW/m^3 [2] for removal of only decay heat (see a solid line of Fig. 6). In case of decay heat and chemical reaction heat (i.e., 0.78 MW), which is generated by graphite and CO oxidations during air ingress, maximum power density decreases down to 4.2 MW/m^3 (see a dotted line of Fig. 6). Such a decrease

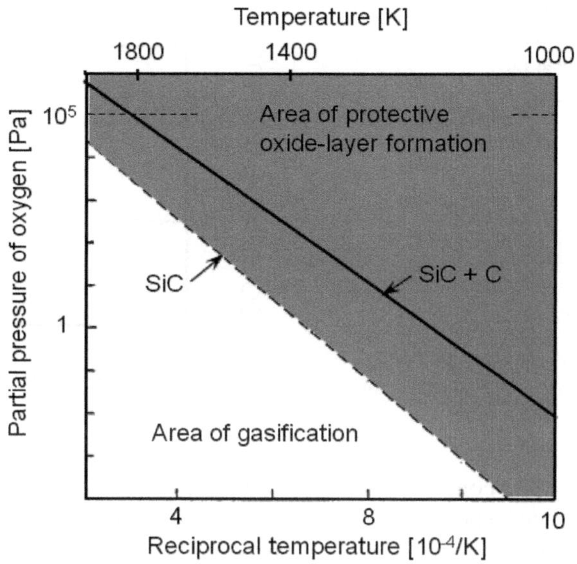

Fig. 7 Formation of protective oxide-layer during oxidation of SiC

of power density may be economically acceptable. Here, power densities of 5.3 MW/m^3 and 4.2 MW/m^3 correspond to thermal powers of 600 MW and 480 MW thermal powers respectively.

It is found that not only low power density but also large heat capacity of core are required in design in order to remove decay heat from the outer surface of reactor pressure vessel with only thermal radiation and natural convection of the atmosphere. It is also found that if power density is reduced from 5.3 MW/m^3 to 4.2 MW/m^3, the design of HTGR can be made feasible even if heat generated by chemical reactions during air ingress accident is added to decay heat. Further analysis indicates that a smaller body which generates heat is easier to be cooled with only thermal radiation and natural convection. Diameters of rector core, fuel element, fuel compact and fuel particle are 5.5 m, 0.36 m, 0.036 m, and 0.001 m respectively. Consequently, a fuel particle coming apart is most easily cooled.

Figure 7 shows area theoretically possible of formation of oxide-layer but not gasification of SiC while SiC for fuel cladding of HTGR is oxidized due to air ingress accident. Thus, formation of oxide-layer due to oxidation of SiC can restrain oxidation of SiC cladding material.

Figure. 8 shows change in CO concentration and reaction-heat rate with graphite temperature during air ingress accident [8]. CO and reaction-heat rate due to graphite and CO oxidations increase with graphite temperature increase, and then CO and reaction-heat rate decrease as graphite temperature increase for the following reasons:

- Increase of the CO oxidation rate with graphite temperature increase
- Decrease of flow rate of air due to dynamic-viscosity change with gas temperature increase

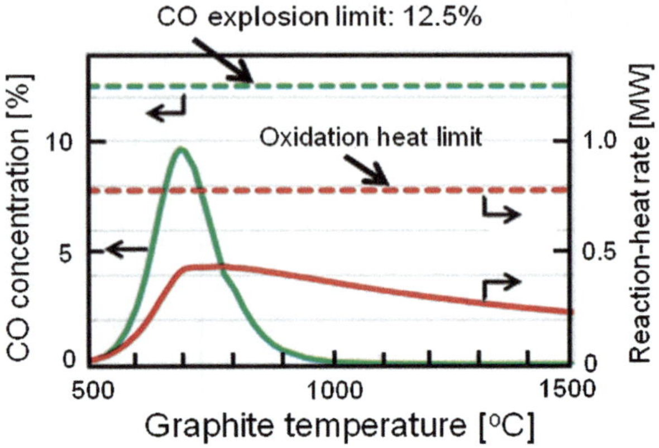

Fig. 8 Change in CO concentration and reaction-heat rate

CO concentration and reaction-heat rate do not exceed CO explosion limit of 12.5% and oxidation heat limit of 0.78 MW used in Fig. 6 respectively. Consequently, counter physical phenomena of oxidation of CO can restrain explosion of CO and air gas-mixture.

4 Discussion

In the F1 accident, there was widespread release of radioactive materials into air, land and sea. The release was so heavy that the accident was on the maximum Level 7 of the International Nuclear Event Scale. Such a tragic accident should never be caused again. So far, Japan has been always striving to overcome failures making the best use of technologies, according to the Japanese philosophy of the techno nationalism. Fortunately, we have already had a technology option capable of overcoming the failures in JAEA. It is a HTGR. HTGR potentially has remarkably high safety. However, so far HTGR development has been focused on economical multi-use of nuclear heat at high temperature. The present concept on HTGR is intended to pursue supreme inherent safety.

In the present regulation low in Japan, countermeasures after severe accident happened are not regulated, but are left to electricity industry sectors. In newly modified law, implementation of countermeasures after DEC will be regulated to safely protect the public and environment. Nuclear power plants have lost reliance of the public because of the F1 accident. Therefore, safety characteristics of HTGR have to be more thoroughly pursued in order to clear regulations modified and to obtain reliance of the public after the F1 accident. So, rough feasibility of the present concept on Inherently-safe HTGR has been just indicated as described in the previous chapters. Experimental and numerical more detailed analysis on inherent safety of HTGR will be conducted by making the most of HTTR.

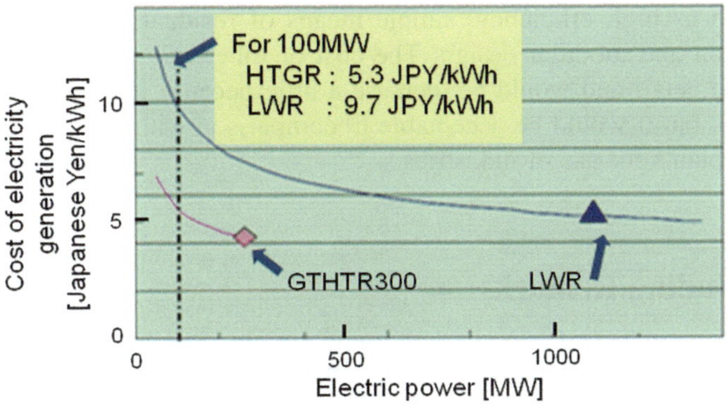

Fig. 9 Cost of electricity generation in GTHTR300

Water ingress into core of HTGR has strong impact. HTGR can be designed so as not to use water as a coolant. While submergence due to external events such as Tsunami and flood might happen at a site of HTGR, nuclear power plants may not be constructed at areas extremely suffering external events, such as a zone of active faulting (earthquake), a seaside area (Tsunami), a dry riverbed (flood, big downpour) and so on; otherwise nuclear power plants should take precautionary measures against the severe external events.

Inherently-safe HTGR does not require any active and/or passive devices and facilities to protect the public and environment, even if HTGR falls into DEC. Two kind of safety designs have to be conducted for two purposes; to secure integrity of a nuclear power plant itself and to protect the public and environment. Needless to say, safety devices and facilities are needed to secure integrity of a nuclear power plant even if inherent safety does not require any safety devices and facilities. Degree of the safety in design will be evaluated with risk assessment for management of nuclear power plants including Inherently-safe HTGR too. High economy is also important in nuclear power plants which have enough safety. Takei et al. [9] reported that the cost of GTHTR300 [2] is about 4 Japanese Yen per kWh as the same level as that of LWR as shown in Fig. 9. The cost was evaluated under the same conditions as those in the cost evaluation of LWR [9]. Thus, HTGR has a competitive economics even a small-sized nuclear power plant. The reason of this economical competitiveness of HTGR is comprehensible in case that the cost of a reactor system would account for mere around 2% of the total cost of a nuclear power plant consisting of a reactor system, a cooling system, an auxiliary facility system of reactor and plant, a measurement/control system, an electricity system, a turbine/generator system, a fuel treatment system, a waste disposal system, buildings, and so on, even though the power density of HTGR (about 5 MW/m^3) is one order of magnitude less than that of LWR (about 50 MW/m^3). The one tenth power density of the HTGR would magnifies the reactor cost roughly ten times which is about 20% (equal to ten times of the 2% described above), but the excess of the reactor cost could be sufficiently offset by the unique design features of the

HTGR such as high efficiency, simple means of residual heat removal, system simplification and modular design. The cost of Inherently-safe HTGR should be evaluated in detail and would be increase a little because inherently safe features are pursued, but it would be acceptable in comparison with those of other energy generation plants for use of industries.

5 Concluding Remarks

To obtain public reliance on nuclear safety in a post Fukushima-crisis Japan, even if any unexpected accidents happen and nuclear power plant falls into DEC, so as not to extremely affect the public and environment, Inherently-safe HTGR is one of the most promising solutions to protect the public and environment because it can attain stable state and retain FP within cladding material only with counter physical phenomena but not any active and/or passive devices and facilities. It has been already verified by using HTTR that a part of the counter physical phenomena such as Doppler effect, thermal radiation and natural convection of the atmosphere actually work to restrain cause events. Design analysis also confirmed that feasible design parameters can be determined for all of counter physical phenomena to work to restrain cause events.

References

1. Kunitomi K, Yan X, Nishihara T, Sakaba N, Mouri T (2007) JAEA's VHTR for hydrogen and electricity cogeneration: GTHTR300C. Nucl Eng Technol 39–1:9–20
2. Yan X, Kunitomi K, Nakada T, Shiozawa S (2003) GTHTR300 design and development. Nucl Eng Des 222:247–262
3. Saito S et al (1994) Design of high temperature engineering test reactor (HTTR), JAERI 1332. Japan Atomic Energy Research Institute, Ibaraki
4. Fujikawa S, Hayashi H, Nakazawa T, Kawasaki K, Iyoku T, Nakagawa S, Sakaba N (2004) J Nucl Sci Technol 41:1245–1254
5. Takamatsu K, Ogawa M et al (2011) High-temperature continuous operation of the HTTR. Trans Atomic Energy Soc Jpn 10–4:290–300
6. Takamatsu K et al (2012) Reactor kinetics in a loss-of-forced-cooling (LOFC) test of HTGRs. In: Proceedings of the 20th International Conference on Nuclear Engineering (ICONE-20-54641 presented Aug. 1, 2012)
7. Sato H, Ohashi H, Tachibana Y, Kunitomi K, Ogawa M (2012) Analysis of core heat removal capability under DLOFC accidents for HTGRs. In: Proceedings of the 6th International Topical Meeting on High Temperature Reactor Technology "HTR2012": HTR2012-8-003 presented Nov. 1, 2012.
8. Ohashi H, Sato H, Tachibana Y, Kunitomi K, Ogawa M (2012) Feasibility study on inherently-safe HTGR for air ingress accident. In Proceedings of the 6th International Topical Meeting on High Temperature Reactor Technology "HTR2012": HTR2012-8-004 presented Oct. 31, 2012.
9. Takei M, Kosugiyama S, Mouri T, Katanishi S, Kunitomi K (2006) Economical evaluation on gas turbine high temperature reactor 300 (GTHTR300). J Nucl Sci Technol 5–2:109–117

(ii)
Session Papers

Monte Carlo Calculations of γ-Rays Angular Distribution Scattering from ^{11}B in (γ, γ) Interaction

Hani Negm, Mohamed Omer, Ryota Kinjo, Yong Woon Choi, Kyohei Yoshida, Torgasin Konstantin, Marie Shibata, Kyohei Shimahashi, Hidekazu Imon, Heishun Zen, Toshitada Hori, Toshiteru Kii, Kai Masuda, and Hideaki Ohgaki

Abstract An investigation of angular distribution of scattered gamma-rays is important to get information about an efficient arrangement of γ-ray detectors and it is necessary to design a real detection system of the inspection system by using a (γ, γ) interaction. Angular distribution of scattered gamma radiation from ^{11}B at 4,440 keV from transition J^{π} ($5/2^{-} \rightarrow 3/2^{-}$) has been simulated by extended GEANT4. In the simulation, seven $LaBr_3$:Ce detectors were recording the scattered photons from nuclear resonance fluorescence (NRF) process in a plane perpendicular to the incident polarized γ-ray beam. The γ-ray beam was assumed to be monoenergetic and linearly polarized with energy spread of 5%. All the $LaBr_3$:Ce detectors were similar in the diameter of 1.5 in. and length of 3 in., positioned at seven different directions. Angular distribution of the scattered γ-rays is discussed in terms of the detectors' positions with respect to the target and incident γ-ray beam. The result, which indicates the largest count rate from NRF signals is backward (135° and 225°) and forward (45° and 315°) directions with respect to the incident gamma-ray, is useful when using the NRF process in inspection of the special nuclear materials (SNM) like ^{235}U and ^{239}Pu.

Keywords Angular distribution • GEANT4 • $LaBr_3$:Ce • NRF

H. Negm (✉) • M. Omer
Institute of Advanced Energy, Kyoto University, Gokasho, Uji, Kyoto 611-0011, Japan

Physics Department, Faculty of science, Assiut University, Assiut 71516, Egypt
e-mail: haninegm@iae.kyoto-u.ac.jp; negm_sci@aun.edu.eg

R. Kinjo • Y.W. Choi • K. Yoshida • T. Konstantin • M. Shibata • K. Shimahashi • H. Imon
H. Zen • T. Hori • T. Kii • K. Masuda • H. Ohgaki
Institute of Advanced Energy, Kyoto University, Gokasho, Uji, Kyoto 611-0011, Japan

T. Yao (ed.), *Zero-Carbon Energy Kyoto 2012*, Green Energy and Technology,
DOI 10.1007/978-4-431-54264-3_21,

1 Introduction

To utilize the gamma-ray for the active probe of the inspection of hidden special nuclear materials (SNMs) [1], a well-designed detection system is indispensable. For this purpose a numerical simulation code which can treat the nuclear resonance fluorescence (NRF) is required to assist the design work. In addition the angular distribution of outgoing gamma rays depends upon the mode of the photon reaction.

The angular distribution of the emitted fluorescence photons is described by a correlation function, $W(\Theta)$, defined with respect to the direction of the incident photon [2]. The correlation function depends upon the multipolarities of the electromagnetic radiation involved in the emission transition. Participating multipolarities in turn are constrained by selection rules determined by the conservation of the angular momentum and the parity in the emission process [3]. Analytic expressions for the correlation function can be found in e.g. [2, 4].

The angular distribution of the outgoing photon will be sampled from correlation functions, $W(\Theta)$, appropriate to the multipolarities of the electromagnetic radiation. In general, more than one electric and magnetic multipole satisfy the selection rules governing a particular transition and can thus contribute [2], with relative intensities that depend upon the detailed overlap of the initial and final states of nuclear wave functions. Thus the resulting angular distribution is a superposition of the distributions for the contributing multipoles. It is anticipated that in many cases where both electric and magnetic multipoles can contribute to the same transition, the mixing ratio is unknown in most cases.

GEANT4 (GEometry ANd Tracking) [5] is a Monte Carlo software toolkit to simulate the passage of particles through matter. The NRF process was modeled by creating a new NRF class that calculated the interaction cross-section and the nuclear de-excitation data in extended GEANT4 [6] as well, the specific angular distribution of its particular transition of the nucleus in terms of the type of transition dipole or quadrupole transition, in case of our target the transition is dipole M1.

In this paper we will describe the fundamental equation of the angular distribution of the resonantly scattered gamma-rays in the NRF process, which is taken into account of the simulation code GEANT4. To investigate the propagation of scattered gamma-rays from NRF process and determine the angular distribution of gamma-rays around the target of Boron-11 at 4,440 keV in the plane perpendicular to the incident direction of the polarized gamma-ray beam, the GEANT4 toolkit has been used.

2 Angular Distribution Calculation

The angular correlation function $W(\Theta)$ of scattered photon γ_2 with respect to the incoming photon γ_1 (beam), in cases $L \leq 2$, can be written as:

$$W(\Theta) = \sum_{v=0,2,4} A_v\ (1)^{\bullet}A_v\ (2)^{\bullet}P_v(\cos\Theta) \tag{1}$$

The angular distribution of photons scattered off an even–even nucleus through dipole transitions (spin sequence 0-1-0) in case of using non-polarized gamma ray is given by:

$$W(\Theta)_{Dipole} = \frac{3}{4}{}^{\bullet}(1 + cos^2\Theta) \tag{2}$$

and through quadrupole transitions (spin sequence 0-2-0) is given by:

$$W(\Theta)_{Quadrupole} = \frac{5}{4}{}^{\bullet}(1 - 3\ cos^2\ \Theta + 4\ cos^4\ \Theta\ ) \tag{3}$$

Angular distribution of gamma rays in case of using linearly polarized beam is given by:

– In case of dipole

$$W(\Theta,\ \phi)_{Dipole} = W(\Theta)_{Dipole} \mp \frac{3}{4}{}^{\bullet}\left(1 - cos^2\Theta\right)\cos 2\phi \tag{4}$$

where minus sign for electric dipole (E1), and plus sign for magnetic dipole (M1).
– In case of quadrupole

$$W(\Theta, \phi)_{Quadrupole} = W(\Theta)_{Quadrupole} \pm \frac{5}{4}{}^{\bullet}\left(1 - 5cos^2\ \Theta + 4\ cos^4\ \Theta\ \right)cos2\phi \tag{5}$$

In general case, when L = 1 dipole transition, when there is no mixing ratio of E2 (E2/M1, mixing ration = 0).

$$W(\Theta,\ \phi)_{Dipole} = 1 + (F_2)^2\ [p_2 \mp \frac{1}{2}\ \cos 2\phi^{\bullet}P_2^{(2)}] \tag{6}$$

where:

$$P_2 = \frac{1}{2}(3\ cos^2\Theta - 1)$$

$$P_2^{(2)} = 3(1 - cos^2\Theta)$$

The F-coefficients can be found in various compilations (e.g., Siegbahn 1965 [7]; Wapstra et al. 1959 [8]).

In our case, the angular distribution of ^{11}B of transition (5/2 → 3/2), which considered as M1 transition is given by:

$$W(\Theta, \phi)_{Dipole} = \frac{7}{100}\left(3\ cos^2\Theta + \frac{93}{7}\right) \mp \frac{21}{100}\cos 2\phi\ (1 - cos^2\Theta) \tag{7}$$

All calculation of angular distribution which represent a part of the NRF process is made by the GEANT4 toolkit which extended to NRF processes.

The simulation is performed under the assumption that the bulk density of the B_4C is 2.52 g/cm^3, which ^{11}B represent 64 wt.%. The target shape is assumed as box with $5.08 \times 5.08 \times 5.08$ cm^3. The energy of incident gamma ray is assumed to have Gaussian distribution with energy spread (full width at half maximum, FWHM) 5%. The number of incident gamma-rays irradiated the target is 2×10^9, the beam width of assumed to be 19.05 mm, and 100% linearly polarized gamma-ray beam in vertical plane.

3 Results and Discussion

The NRF signals were recorded by $LaBr_3$:Ce detectors at each angle from 45° to 315° by step 45°. It is obvious that we cannot use the 0° where strong transmission gamma-rays will hit the detector. The results are shown in Figs. 1b and 3 summarized all results of NRF signals from all detectors at different distances from the center of target which setting in Fig. 1a, which illustrates the distribution of propagation of gamma-rays from the target after NRF process and we can see from this figure the distribution not isotropic around the target (not homogeneous in all direction) but it has some distribution dependent on the angles with respect to the incident direction of polarized gamma-ray or by another word it is the angular distribution. Also, the distribution has a shape little bit like a cosine or sine shape around the target as well as the distribution shape has symmetry as mirror at 180°. It means the angular distribution is symmetric around the incident beam.

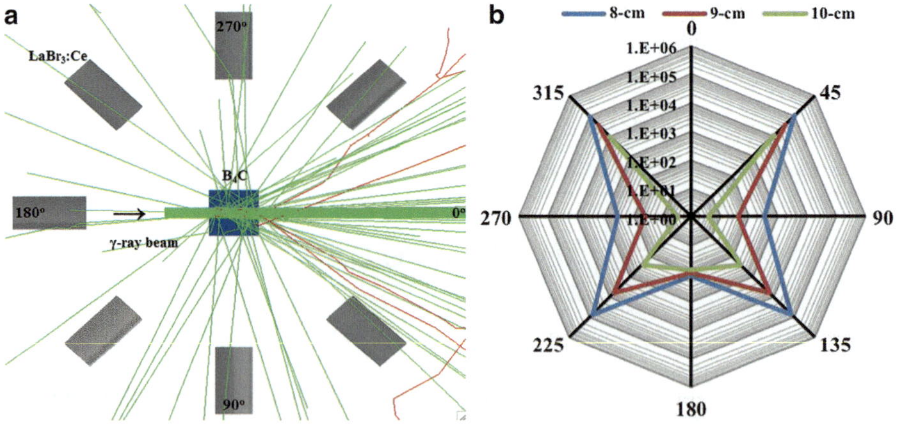

Fig. 1 (**a**) The schematic view of geometry configuration of detectors to record the NRF signals (from (γ, γ) interaction) in seven directions around the target (B_4C) and perpendicular to the direction of polarization of gamma. (**b**) Shows the distribution around the target at different distances

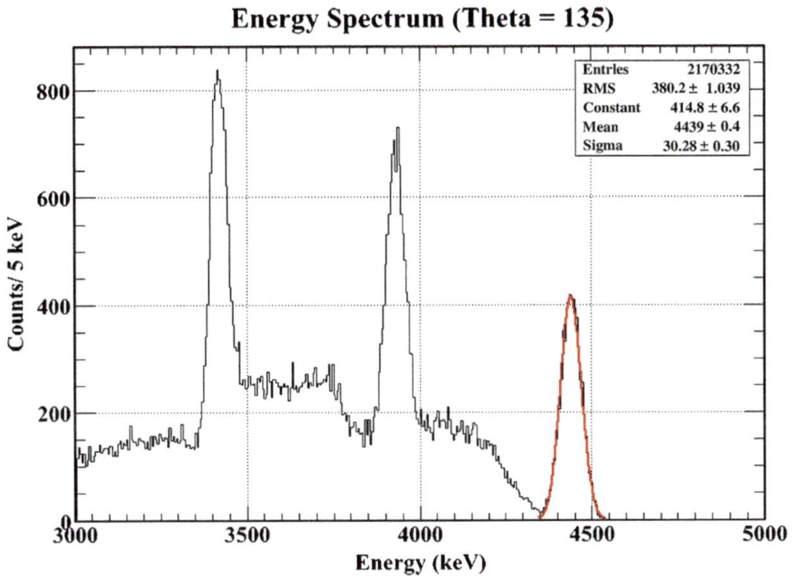

Fig. 2 The energy spectrum of NRF signal from one detector at 135°, and 9 cm from center of target by extended GEANT4, the *red line* represent the Gaussian fitting to NRF signal

So, we can predict from the calculation the angular distribution of (γ, γ) interaction as well as the best and the optimum setting for the measuring system of NRF experiments. Figure 2 shows the sample from energy spectrums at 135° at 9 cm of ^{11}B at 4,440 keV transition as shown with the red color line, which shows the clear NRF signal at specific energy, also appears two peaks of single and double escape peaks at 3,929 and 3,418 keV, respectively. Figure 3 illustrates the angular distribution of NRF gamma-rays from ^{11}B target. It should be noted that the detector located at 180° shows slightly different manner, almost the same amount of counts are recorded in three different distances. This is due to the backward scattering gamma-rays from the target. Therefore, we cannot use the 180° angle for setting up the detector in the NRF experiment.

4 Conclusion

Investigation of the gamma rays propagation at different angles, and different distances was performed by an extended GEANT4. The results illustrate the angular dependent distribution of gamma-ray and the largest count rate from NRF signals is backward (135° and 225°) and forward (45° and 315°) directions with respect to the incident gamma-ray. So, the best configuration of the detectors in this case in backward or forward or both to record and detect high count rate of NRF

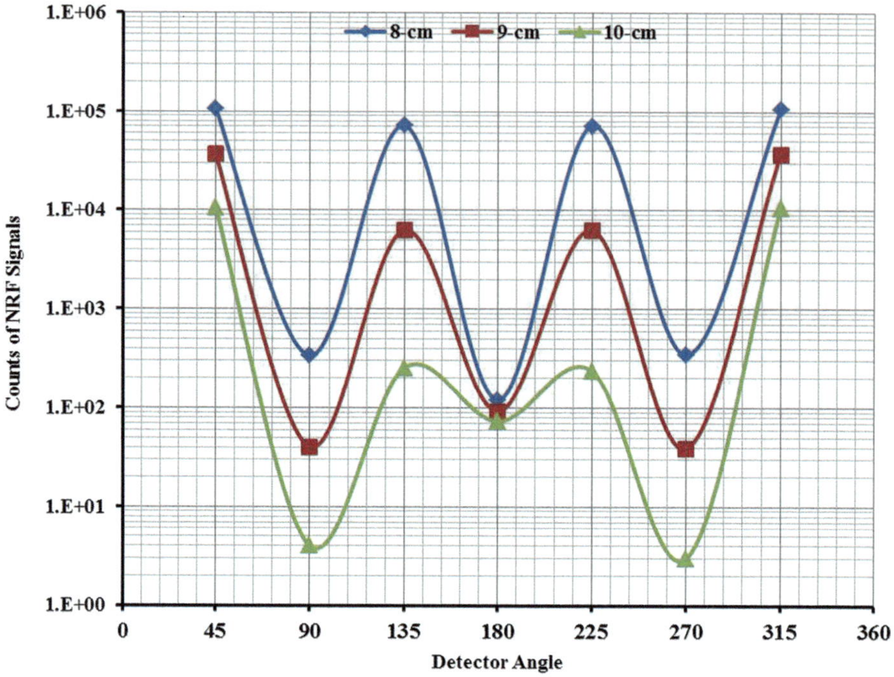

Fig. 3 The results of counts of NRF signal of seven detectors at different angles from 45° to 315°, at three distances at 8, 9, and 10 cm

signals in these directions. This method is important to investigate the angular distribution of special nuclear materials (SNM) like ^{235}U, and ^{239}Pu hidden in the cargo where much complex processes, atomic scattering from the target and surrounding materials could be expected.

Acknowledgment The authors were grateful for the partial support of G-COE program "Energy Science in the Age of Global Warming" Kyoto University. This work is supported by Special Coordination Funds for Promoting Science and Technology in Japan (Grant No. 066).

References

1. Kikuzawa N et al (2009) Nondestructive detection of heavily shielded materials by using nuclear resonance fluorescence with a laser-compton scattering γ-ray source. Appl Phys Express 2:036502
2. Kniessl U et al (1996) Investigation of nuclear structure by resonance fluorescence scattering. Prog Part Nucl Phys 37:349
3. Krane KS (1988) Introductory nuclear physics. Wiley, New York
4. Fagg LW, Hanna SS (1959) Polarization measurements on nuclear gamma rays. Rev Mod Phys 2:711–758

5. Agostinelli S et al (2003) Geant4—a simulation toolkit. Nucl Instrum Methods Phys Res A 506:250
6. Hayakawa T et al (2011) Development of simulation code for NDA using nuclear resonance fluorescence with laser compton scattering gamma-rays. INMM 52th Annual Meeting Proceeding, July 17–21, 2011. Desert Springs JW Marriott Resort, Palm Desert, California, USA
7. Siegbahn K (1965) α, β, γ - Spectroscopy. North Holland, Amsterdam
8. Wapstra AH, Nigh GJ, van Lieshout R (1959) Nuclear Spectroscopy Tables. North Holland, Amsterdam

Optimization of the New Designed FEL Beam Transport Line

Kyohei Yoshida, Heishun Zen, Kensuke Okumura, Kyohei Shimahashi, Marie Shibata, Takuya Komai, Hidekazu Imon, Hani Negm, Mohamed Omer, Yong-Woon Choi, Ryota Kinjo, Toshiteru Kii, Kai Masuda, and Hideaki Ohgaki

Abstract A mid-infrared free electron laser (MIR-FEL) (target wavelength: 5 ~ 20 μm) facility named KU-FEL (Kyoto University Free Electron Laser) was constructed to aid various energy science researches at the Institute of Advanced Energy, Kyoto University. In December 2011, KU-FEL was upgraded by replacing its undulator and optical cavity mirrors. By this upgrade, the tunable range of KU-FEL was improved to 5.5 ~ 15 μm. According to replacing the cavity mirrors, size and divergence of FEL beam at the emitting point was changed. Therefore, we designed and constructed a new MIR-FEL transport line. By using calculation code Zemax (http://www.zemax.com), the condition for keeping the FEL beam radius less than 25 mm during the transportation length of 24 m was determined. In addition, the FEL intensity profile was measured after passing through the constructed transport line. The beam waist at the out-coupling hole was calculated from the measured FEL intensity profile. By using the beam waist calculated from the measurement, we confirmed the validity of the calculated optimum condition.

Keywords FEL intensity profile • Free Electron Laser • Laser science • MIR-FEL

1 Introduction

A mid-infrared free electron laser (MIR-FEL) (target wavelength: 5 ~ 20 μm) facility named KU-FEL (Kyoto University Free Electron Laser) was constructed to aid various energy science researches at the Institute of Advanced Energy,

K. Yoshida (✉) • H. Zen • K. Okumura • K. Shimahashi • M. Shibata • T. Komai • H. Imon
H. Negm • M. Omer • Y.-W. Choi • R. Kinjo • T. Kii • K. Masuda • H. Ohgaki
Institute of Advanced Energy, Kyoto University, Gokasho, Uji-shi, Kyoto, Japan
e-mail: k-yoshi@iae.kyoto-u.ac.jp

T. Yao (ed.), *Zero-Carbon Energy Kyoto 2012*, Green Energy and Technology,
DOI 10.1007/978-4-431-54264-3_22,

Kyoto University. MIR-FEL has excellent features such as widely tunable wavelength, short pulse and high peak power. One of the most interesting applications of MIR-FEL is selective excitation of chemical bond in gas and solid by tuning the FEL wavelength to those vibration frequencies in MIR-region [1]. In KU-FEL, the first laser power saturation at 13.2 μm was achieved in May 2008 [2]. In December 2011, KU-FEL was upgraded by replacing the undulator and the re-designing optical cavity. After the upgrade, the tunable wavelength of KU-FEL was improved from 10 ~ 13 to 5.5 ~ 15 μm.

FEL is the laser amplified by high energy electron beam, which also generates undesired X-ray, γ-ray and neutron. Therefore we cannot freely access to the FEL emitting point, in our case the coupling hole, during FEL lasing. In general, FEL beam has to be transported from emitting point to user station for practical use. The beam transport system developed for MIR-FEL in Science University of Tokyo uses elliptical and parabolic mirrors [3]. The system can make the FEL beam be parallel. However, a demerit of this transport system is the difficulty of alignment. In other facilities such as FOM-FELIX [4] in Netherlands and FELI in Japan, a spherical mirror is used to make FEL beam be quasi-parallel [5]. A merit of this system is easiness of alignment. However, some aberration happens in the transported laser beam.

In our facility, the spherical mirror was adopted for FEL beam transport line [2]. The FEL beam was extracted from the previous optical cavity through the out-coupling hole whose diameter was 2 mm. However, when the FEL facility was upgraded, the diameter of the out-coupling hole was changed to 1 mm to reduce the optical loss of the cavity at short wavelength of FEL. Then, the divergence of extracted FEL beam became almost double, and then the transport efficiency of the FEL beam was decreased because the diameter of the transported FEL beam was larger than the diameter of the mirrors used in the transport line. Therefore, we designed and constructed a new FEL beam transport line for the upgraded FEL facility by changing the position and the radius of curvature of the spherical mirror. The concept of the design of the new FEL beam transport line is to make FEL beam be quasi-parallel with keeping the beam radius, which is defined by two times of σ (standard deviation) of Gaussian beam, less than 25 mm. Because, the diameter of mirrors used in FEL beam transport line in KU-FEL is 50 mm.

The optimum position and the radius of curvature of the spherical mirror were deduced by using the method reported in [6] and software Zemax [7]. The objective of the calculation is to seek the optimum condition which the FEL beam radius less than 25 mm is kept during the FEL beam transport line.

After designing the new FEL beam transport line, we measured the FEL intensity profile at the user station of KU-FEL to calculate the beam waist at out-coupling hole. By using the beam waist calculated from measurement, the optimum condition of the FEL beam transport line was calculated and validated.

2 Design of FEL Transport Line

The evolution of the FEL beam size after reflecting the spherical mirror should be calculated under the condition of different wavelength of FEL in order to determine the optimum position and focal length of spherical mirror. For the calculation, an initial condition of the FEL beam size and divergence at the out-coupling hole should be given. The FEL beam profile from the out-coupling hole is considered to be similar to Airy disk pattern from a circular aperture. We approximated the central part of Airy disk with Gaussian beam whose waist size was estimated to be 0.38 mm regardless the wavelength when the aperture size is 1 mm. We show the configuration of the optics in the FEL beam transport line in Fig. 1. A Mercury Cadmium Zinc Telluride (MCZT) detector is used for observation of power evaluation of FEL.

The beam size of the transported FEL depends on the position and the radius of curvature of the spherical mirror. For the calculation of the beam size at user stations, we defined *F*, and *d*, which are the focal length of the spherical mirror, and the distance between out-coupling hole and spherical mirror, respectively. At first, we investigated the optimum focal length of the spherical mirror. The procedure of the calculation is shown in Fig. 2. The relation between the beam waist at the out-coupling hole, ω_0, and the beam waist made by the reflection at spherical mirror, ω_0', is expressed by Eq. (1) by using *d* and *F* [6]. λ in Eq. (1) is the FEL wavelength. D, which is the distance between the position of the beam waist after the spherical mirror and the surface of the spherical mirror, is shown by Eq. (2).

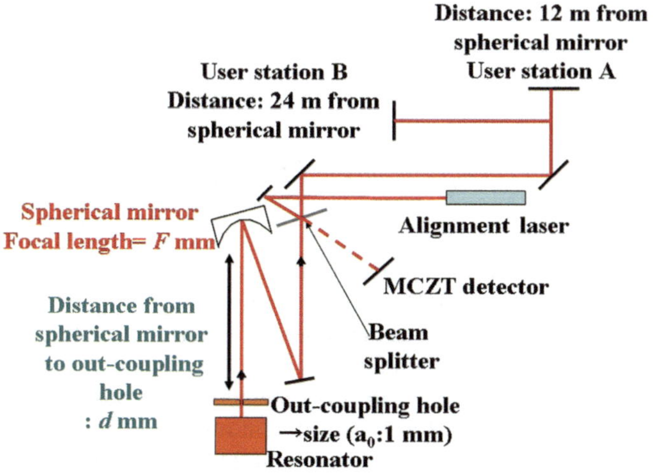

Fig. 1 Outline of the FEL beam transport line

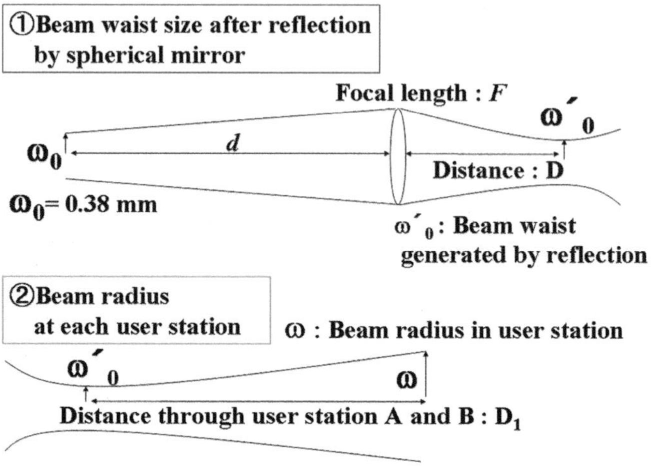

Fig. 2 Procedure of calculation of the spherical mirror position and focal length dependence of the beam radius in each user station

$$\omega_0{}' = \omega_0{}^2 \frac{F^2}{(F-d)^2 + \left(\frac{\pi \omega_0{}^2}{\lambda}\right)^2} \tag{1}$$

$$D = F + (d - F)\left(\frac{\omega_0{}'}{\omega_0}\right)^2 \tag{2}$$

In the second step, ω, which is the beam radius at a particular point in the FEL transport line, is calculated by Eq. (3) by using $\omega_0{}'$ and D_1. D_1 is the distance between the position of the beam waist made by reflection of spherical mirror and user station A and B.

$$\omega = \omega_0{}' \left\{ 1 + \left(\frac{\lambda D_1}{\pi \omega_0{}'^2}\right)^2 \right\}^{\frac{1}{2}} \tag{3}$$

The distances between the spherical mirror and the user station A and B, i.e. $D_1 + D$, are fixed at 12 m and 24 m, respectively.

Figure 3 shows the beam radius dependence on mirror curvature at user stations A and B when the FEL wavelength is 5 μm, 12 μm, and 20 μm (Fig. 3). The horizontal axis in the figure shows the displacement of the position of the spherical mirror from focal length, i.e. *d-F*. Five different mirrors whose focal lengths were 500 mm, 750 mm, 1,000 mm, 1,500 mm, and 2,000 mm were examined. As the results, the beam radius at user station A and B is close to the minimum in any

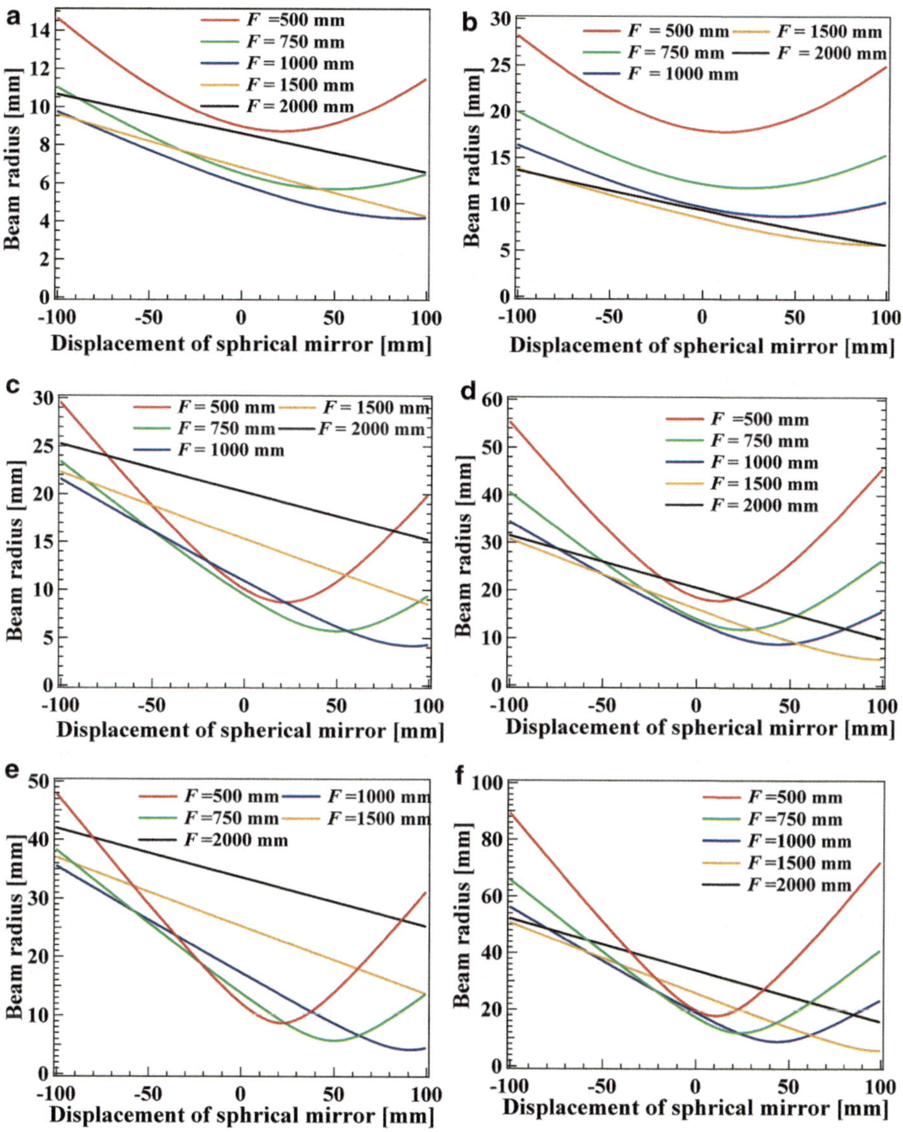

Fig. 3 Spherical mirror position and focal length effect on the beam radius in each user station. 0 mm in x-axis means the distance between the out-coupling hole and spherical mirror is *F* mm. (**a**) Beam radius in user station A when the wavelength is 5 μm. (**b**) Beam radius in user station B when the wavelength is 5 μm. (**c**) Beam radius in user station A when the wavelength is 12 μm. (**d**) Beam radius in user station B when the wavelength is 12 μm. (**e**) Beam radius in user station A when the wavelength is 20 μm. (**f**) Beam radius in user station B when the wavelength is 20 μm

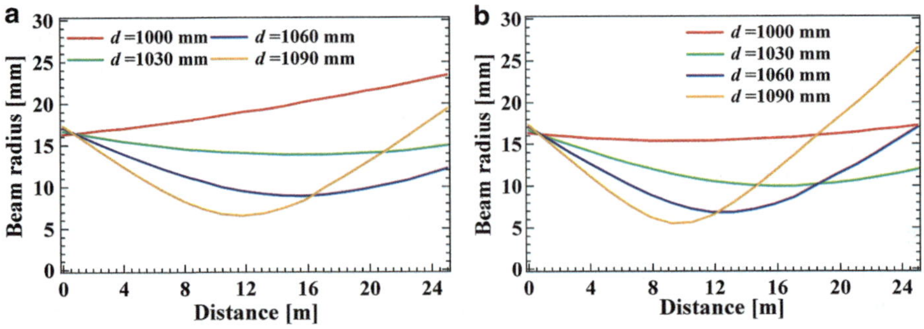

Fig. 4 Distance dependence of beam radius from spherical mirror at 20 μm. (**a**) Horizontal direction. (**b**) Vertical direction

wavelength by using the 1,000 mm mirror. For the efficient FEL beam transport, the smaller beam radius is better, because we can avoid the beam loss due to cut by duct of the FEL transport line. Therefore, the 1,000 mm mirror is the best among five candidates for the FEL beam transport line in KU-FEL.

In the third step, we investigated the optimum position of the spherical mirror. An optics calculation software, Zemax was used in this calculation. In this calculation, the FEL wavelength was set to the longest wavelength of our target, 20 μm, which has the largest beam broadening by diffraction. The results are shown in Fig. 4. The horizontal axis of Fig. 4 shows the distance from spherical mirror. The 12 m and 24 m at horizontal axis in Fig. 4 represent the position at the user station A and B, respectively. From these figures, we found that the beam radius in the FEL beam transport line is not over 25 mm when the spherical mirror is located at the points, $d = 1{,}030$ mm and $d = 1{,}060$ mm. In the case of $d = 1{,}060$ mm, the difference of the beam radius at user station A and B is larger than the case of $d = 1{,}030$ mm. It means the divergence of the case $d = 1{,}060$ mm is larger than the case of $d = 1{,}030$ mm. Therefore, the best condition of the FEL beam transport line in KU-FEL is to use the 1,000 mm mirror and locate at $d = 1{,}030$ mm.

3 Confirmation of the Beam Waist at Out-Coupling Hole

In the calculation of the previous chapter, the initial condition of the FEL beam at the out-coupling hole was hypothecated as the Gaussian beam which beam waist is 0.38 mm in any wavelength. For the practical usage of the beam transport line, it is important to investigate the real condition of the FEL beam at the out-coupling hole. Therefore, the FEL intensity profile was measured for confirming beam radius and the beam waist at out-coupling hole. For this experiment, we used the spherical mirror of $F = 750$ mm. Next, the optimum distance between the spherical mirror and the

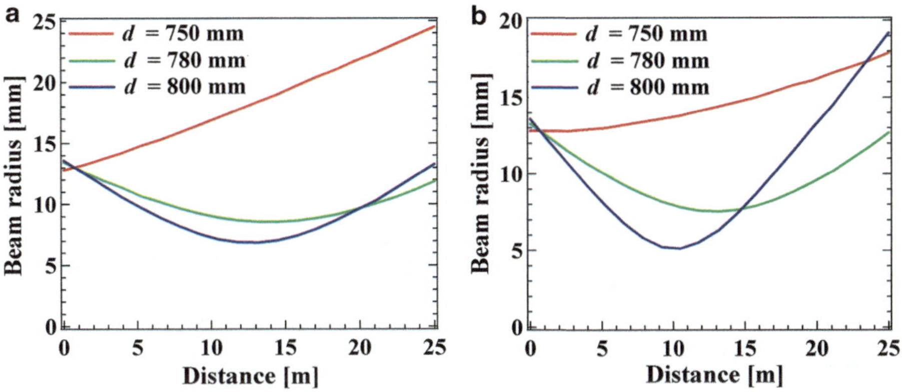

Fig. 5 Dependence on the beam radius of the distance from spherical mirror when the wavelength is 20 μm. (**a**) Horizontal direction. (**b**) Vertical direction

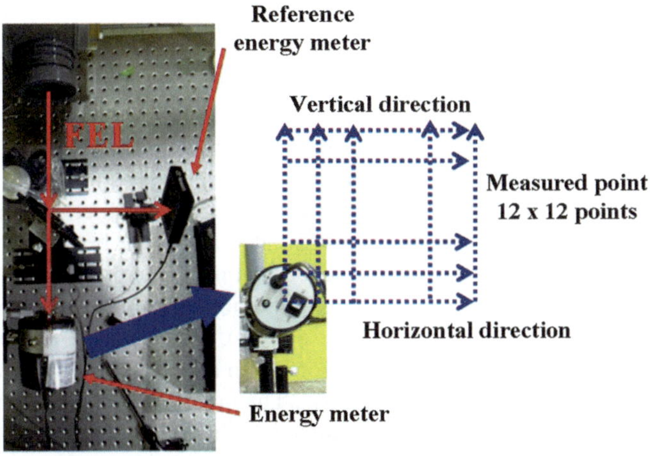

Fig. 6 Measurement system of FEL intensity profile

out-coupling hole, which is *d*, was calculated in the case of the 750 mm mirror. The calculation method is the same as Chap. 3. The results are shown in Fig. 5.

From these figures, in the case of $d = 780$ mm, we found the change of the beam radius during FEL transport line was small. Therefore, the spherical mirror of the 750 mm was located at the point, $d = 780$ mm.

After setting the 750 mm spherical mirror, we measured the FEL intensity profile at user station A and B. The measurement method is shown in Fig. 6. We set a 1 mm pinhole in front of the detector (pyroelectric detector QE8SP-I-BL-BNC), and we measured the intensity in 12×12 points by scanning the detector to horizontal and vertical directions. The measured intensity was normalized to a reference energy

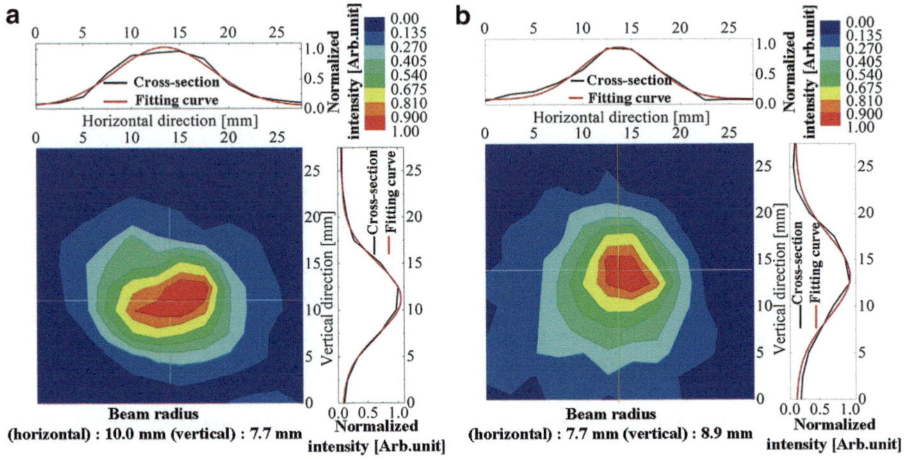

Fig. 7 FEL intensity profile in each user station when the wavelength is 12 μm. (**a**) User station A. (**b**) User station B

meter (pyroelectric energy detector 818E-20-50S) and the results are shown in Fig. 7. The wavelength of the measured FEL intensity profiles is 12 μm. The FEL intensity profile was close to the Gaussian distribution. From the Gaussian fitting, the beam radius in horizontal and vertical direction were deduced as 10.0 mm and 7.7 mm at user station A, and 7.7 mm and 8.9 mm at user station B, respectively.

From the beam radius of the measured FEL intensity profile, we calculated the beam waist at the out-coupling hole. In this calculation, the beam radius is calculated only in vertical since we tilted the spherical mirror in horizontal axis as shown in Fig. 1 and it causes error of focal length and aberration in horizontal direction. The calculation was done by using Eqs. (1), (2), and (3). The definition of the characters is the same as Chap. 3 except for D_1. In this calculation, D_1 was defined the distance between user station B and ω_0'. The ω_0' and D_1 are calculated from ω by using Eq. (3). ω is measured beam radius, which are 7.7 mm and 8.9 mm. From the calculation results, ω_0' was 3.0 mm or 7.6 mm. Moreover, D_1 is 6.5 m when ω_0' is 3.0 mm. D_1 is 9.3 m when ω_0' is 7.6 mm. Then, ω_0 and d are calculated from ω_0' and D by using Eqs. (1) and (2). D can be calculated from D_1 because $D_1 + D$ is 24 m. As the results, ω_0 and d are 0.13 mm and 781 mm when D_1 and ω_0' are 6.5 m and 3.0 mm, or ω_0 and d are 0.28 mm and 769 mm when D_1 and ω_0' are 9.3 m and 7.6 mm.

By using the calculated values of ω_0 and d, the beam radius of vertical direction at user station A and B is calculated by using Zemax. Calculation condition is the same as Chap. 3 except for ω_0 and d. The calculated values ω_0', D_1, ω_0, d, and the beam radius at user station A and B are summarized in the Table 1. When the ω_0 and d are 0.13 mm and 781 mm, the beam radius at user station A and B became 5.2 mm and 12 mm, respectively (solution 1). Next, when the ω_0 and d are 0.28 mm and 769 mm, the beam radius at user station A and B became 7.6 mm and 9.0 mm

Table 1 Summary of the calculated values from the measured beam profile

	ω_0' (mm)	D_1 (m)	ω_0 (mm)	d (mm)	Beam radius at user station A (mm)	Beam radius at user station B (mm)
Solution 1	3.0	6.5	0.13	781	5.2	12.0
Solution 2	7.6	9.3	0.28	769	7.6	9.0

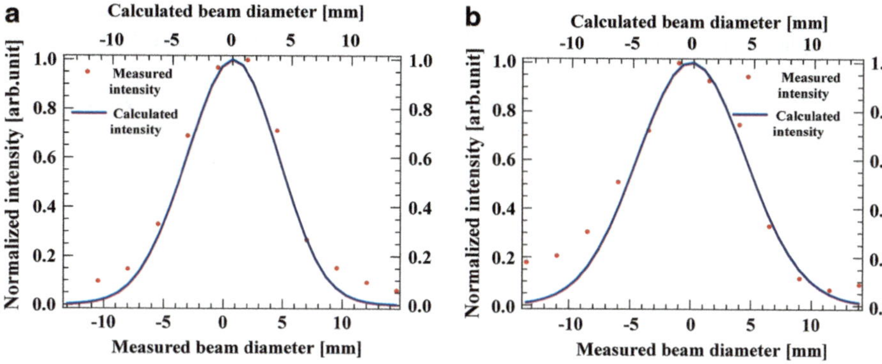

Fig. 8 Comparison between the measured cross-section and the cross-section calculated by using measured value of vertical direction. (**a**) User station A at 12 μm. (**b**) User station B at 12 μm

respectively (solution 2). Solution 2 is consistent with the measured values from FEL intensity profile. In addition, the cross-section of the vertical direction at 12 μm is calculated by using the values of solution 2 by Zemax. The results are shown in Fig. 8. From these figures, the cross-section of vertical direction in measured FEL intensity profile is also consistent with the cross-section calculated by Zemax. Therefore, we could confirm ω_0 is 0.28 mm, and the precious value of d in this experiment is 769 mm.

Finally, by using this confirmed, $\omega_0 = 0.28$ mm, we optimized the focal length and the spherical mirror position at 12 μm. The five mirrors, whose focal lengths were 500 mm, 750 mm, 1,000 mm, 1,500 mm, and 2,000 mm were examined. The results are shown in Fig. 9. From Fig. 9, we could confirm the beam radius was kept less than 25 mm, and the beam radius was also close to the minimum by using the 1,000 mm mirror. Next, the position of the 1,000 mm mirror was calculated. The results are shown in Fig. 10. From Fig. 10, we found the beam diameter was kept less than 25 mm when d is 1,030 mm and 1,060 mm. In addition, the divergence of the beam radius is small in horizontal and vertical direction when d is 1,030 mm.

It is suggested that the optimum position of the 1,000 mm mirror is $d = 1{,}030$ mm from the calculation in the Chap. 3. In addition, we could confirm the optimum position of the 1,000 mm mirror is $d = 1{,}030$ mm from the calculation in Chap. 4. We are preparing to install the 1,000 mm mirror for the optimized FEL beam transport line.

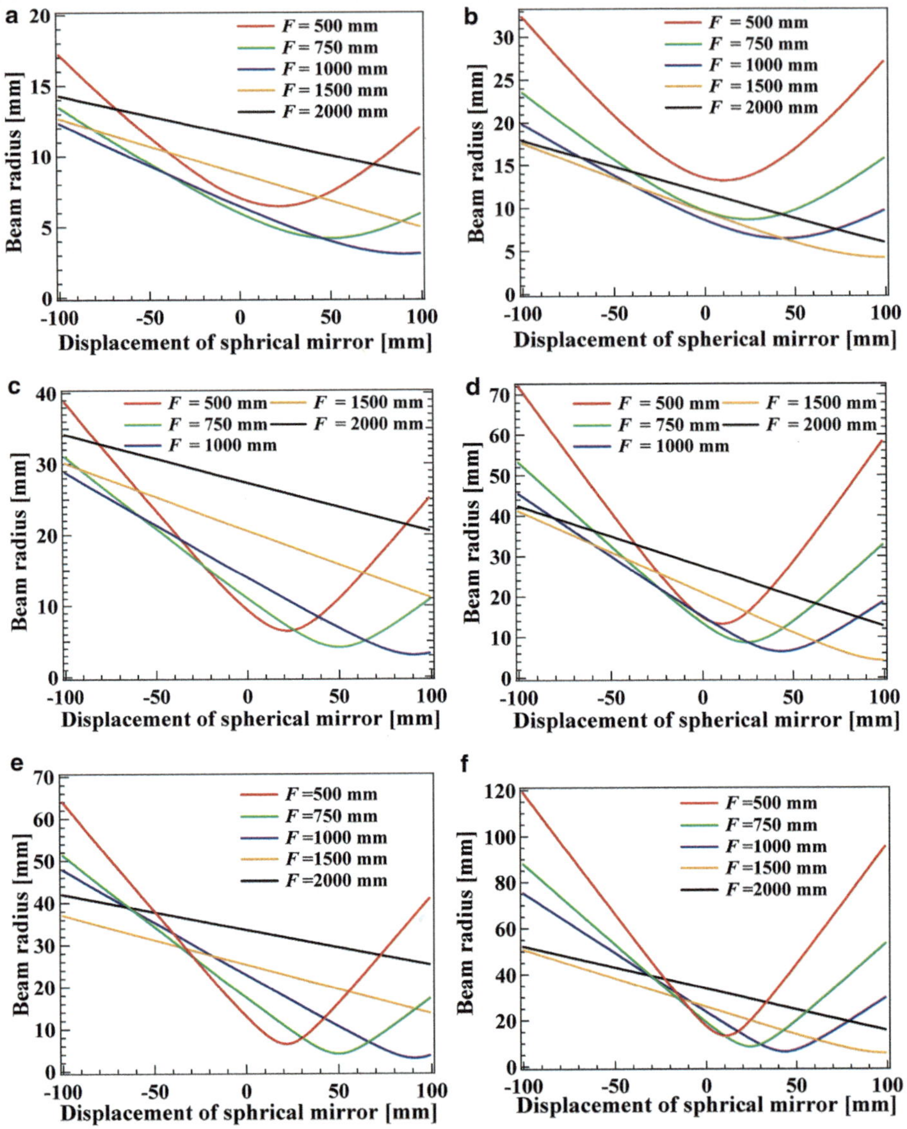

Fig. 9 Spherical mirror position and focal length effect on the beam radius in each user station by using measured value of ω_0. (**a**) Beam radius in user station A when the wavelength is 5 μm. (**b**) Beam radius in user station B when the wavelength is 5 μm. (**c**) Beam radius in user station A when the wavelength is 12 μm. (**d**) Beam radius in user station B when the wavelength is 12 μm. (**e**) Beam radius in user station A when the wavelength is 20 μm. (**f**) Beam radius in user station B when the wavelength is 20 μm

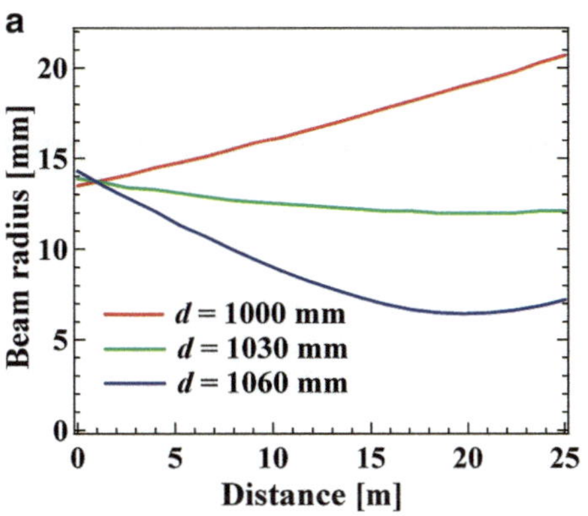

Fig. 10 Dependence on the beam radius of the distance from spherical mirror by using the measured value of ω_0 when the wavelength is 20 μm. (**a**) Horizontal direction. (**b**) Vertical direction

4 Conclusion

We constructed a new FEL beam transport line. Based on the rough estimation of the initial condition of FEL beam, the optimum condition was to put a spherical mirror with a focal length of 1,000 mm at the position $d = 1{,}030$ mm for 20 μm wavelength FEL. To confirm the actual beam parameter of FEL beam, we measured the FEL intensity profile at user stations, and the beam waist at the out-coupling hole was deduced to be 0.28 mm. By using this measured beam waist, the optimum focal length and position of the spherical mirror at 20 μm was calculated to be 1,000 mm and $d = 1{,}030$ mm, respectively.

Acknowledgment This work was supported by the Kyoto University's Global Center of Excellence (GCOE) program of the Graduate School of Energy Science in the Age of Global Warming.

References

1. Tully JC (2006) Mode-selective control of surface reactions. Science 312:1004–1005
2. Ohgaki H et al (2009) Status of the MIR-FEL facility in Kyoto University. Proceedings of FEL, pp 572–575
3. Nomaru K et al (2000) Optical beam transport system at FEL-SUT. Nucl Instrum Methods Phys Res A 445:379–383
4. Oepts D et al (1995) The free-electron-laser user facility FELIX. Infrared Phys Technol 36:297–308

5. Saeki K et al (1995) Optical cavity and transport system for FELI. Nucl Instrum Methods Phys Res A 358:ABS 56–ABS 58
6. Kogelnik H (1965) The image of optical modes-resonators with internal lenses. Bell Syst Tech J 44:455–494
7. Z. Radiant Zemax LLC (2011) *Zemax software for Optical System Design : Getting started using Zemax*, http://www.zemax.com

Diamagnetic Effect on the Ripple-Induced Losses of Energetic Ions in a Non-Axisymmetric Tokamak Plasma

Michinao Bunno and Yuji Nakamura

Abstract In a tokamak plasma, energetic ion losses are increased by the toroidal field (TF) ripple, which is mainly created by the finite number of the toroidal field coils (TFCs). The plasma current alters magnetic field structures and energetic ion orbits. The diamagnetic effect caused by the poloidal plasma current (i.e., the diamagnetic current) is one of the most important finite beta effects. We investigated how the diamagnetic effect affects energetic ion losses using the magnetohydrodynamic (MHD) equilibrium calculation code VMEC and the orbit following Monte Carlo code F3D-OFMC. As a result, it was found that the weak diamagnetic effect increased the fusion alpha particle losses, while the strong diamagnetic effect reduced them.

Keywords Alpha particle • Finite beta effect • Fusion plasma • MHD equilibria • Ripple

1 Introduction

The finite number of the toroidal field coils (TFCs) generates the non-axisymmetric field in a tokamak plasma. This field is called toroidal field (TF) ripple. Even if energetic ions are successfully confined in a axisymmetric tokamak, these might be lost by the TF ripple. Fusion alpha particle losses especially induce several serious issues such as the deterioration of the plasma heating efficiency and the heat load concentration.

In a finite beta plasma, magnetohydrodynamic (MHD) equilibrium currents change the magnetic field structure, and it affects energetic ion losses. This effect is one of the most important finite beta effects. Finite beta effects on energetic ion

M. Bunno (✉) • Y. Nakamura
Graduate School of Energy Science, Kyoto University, Gokasho, Uji, Kyoto, Japan
e-mail: bunno.michinao.44v@st.kyoto-u.ac.jp

T. Yao (ed.), *Zero-Carbon Energy Kyoto 2012*, Green Energy and Technology, DOI 10.1007/978-4-431-54264-3_23,

orbits, such as the Shafranov-shift and the ripple plasma current were clarified in a relatively low beta plasma [1, 2]. While in a high beta plasma, the diamagnetic effect may strongly affect them. The diamagnetic effect, which is induced by the poloidal plasma current (i.e., the diamagnetic current), reduces the vacuum toroidal filed and significantly alters the field strength $|B|$ contours [3]. Since the reflection point of trapped particles follows the $|B|$ contours, the changes of $|B|$ contours affect the banana orbits of energetic ions.

In this study, the diamagnetic effect on the toroidal field was obtained by resolving the MHD equilibrium calculation in a rippled tokamak. To analyze the diamagnetic effect on energetic ion losses, five cases of the three-dimensional (3D) field structures were created by changing the magnitude of this effect.

This paper is organized as follows. In Sect. 2, the calculation model and the calculation code are introduced. The mechanism of diamagnetic effect on energetic ion losses is discussed in Sect. 3. Finally, Sect. 4 summarizes the findings and conclusions of this study.

2 MHD Equilibrium Calculation

First, the two-dimensional (2D) MHD equilibrium field was calculated using the VMEC code which uses the inverse spectral method [4, 5]. The specified pressure and safety factor profile are shown in Fig. 1a–c shows the vacuum ripple ratio δ, which is defined by

$$\delta = \frac{B_{\max} - B_{\max}}{B_{\max} + B_{\max}}, \tag{1}$$

where $B_{\max}$ and $B_{\min}$ are the maximum and minimum field strength along the toroidal direction on the fixed R and Z position. The 18 toroidal field coils were set at the toroidal angles of

$$\phi = \frac{2\pi(i-1)}{18} \qquad (i = 1, 2, 3 \ldots 18). \tag{2}$$

The maximum vacuum ripple ratio was less than 0.01. This value almost equals to that for the ITER. By superimposing the vacuum ripple component on the 2D MHD equilibrium field, the approximated 3D field structure can be created in this study.

Figure 2a shows the vacuum and equilibrium toroidal field B_t along the R direction. The vacuum toroidal field is approximately proportional to $1/R$. Due to the diamagnetic effect, the toroidal field was reduced at the outer torus and the $|B|$ contour curvatures became higher. To analyze the diamagnetic effect on energetic ion losses, five cases of the 3D field structures were created by changing the toroidal field strength generated by the poloidal plasma current. In this study, each field case

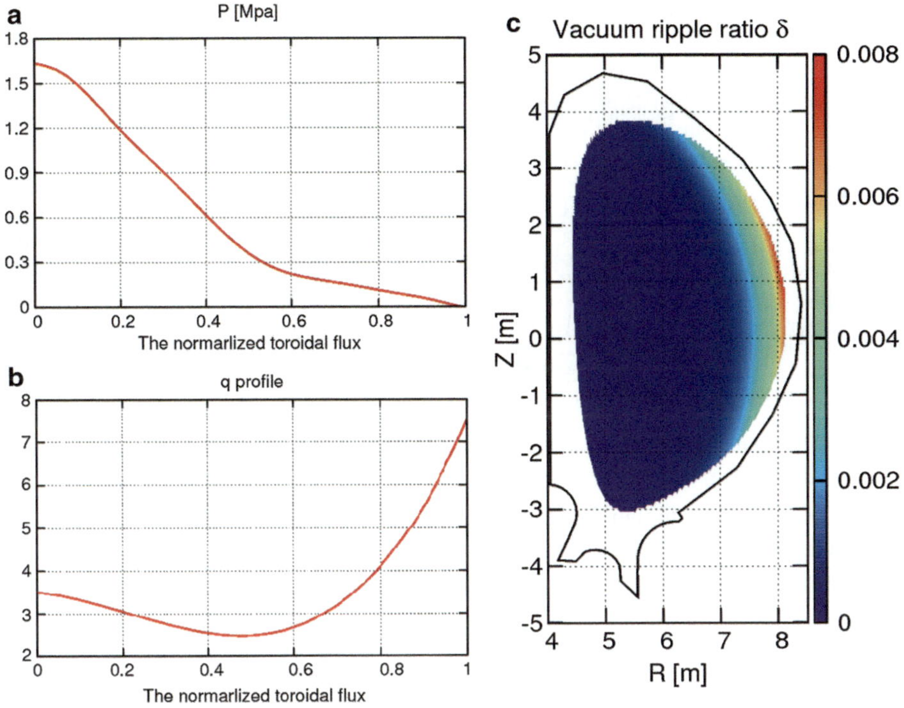

Fig. 1 Specified pressure profile (**a**), the safety factor profile (**b**) and the vacuum ripple ratio (**c**)

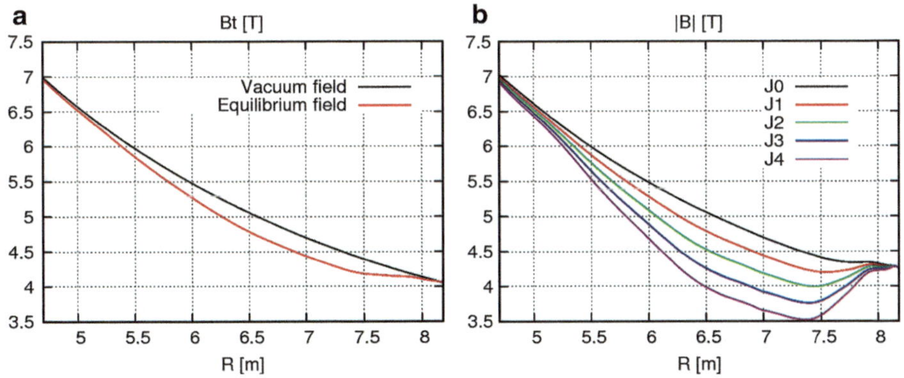

Fig. 2 The vacuum and equilibrium toroidal field strength (**a**) and the field strength $|B|$ (**b**) for each field model along the R direction at $(Z,\varphi) = (0.5\text{ m},0)$

was called J0, J1, J2, J3 and J4, in which the magnitude of the diamagnetic effect was multiplied by 0, 1, 2, 3 and 4, respectively. Figure 2b shows the field strength $|B|$ along the R direction at $(Z,\varphi) = (0.5\text{ m},0)$ for each field case. Figure 3 show the $|B|$ contours at $\varphi = 0$ for each field model. As mentioned above, the contour

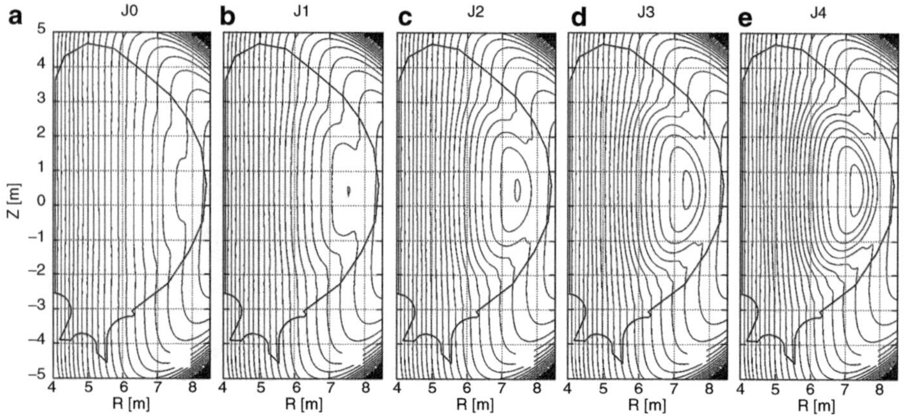

Fig. 3 The contours of the field strength $|B|$ at $\varphi = 0$ for each field model

curvatures became higher and the closed contour region was extended with the increase of the diamagnetic effect. In the next section, we will discuss how the changes of $|B|$ contour affect energetic ion losses.

3 Diamagnetic Effects on the Energetic Ion Losses

Changes of the $|B|$ contours strongly affect energetic ion orbits. When the magnetic moment

$$\mu = \frac{mv^2}{2|B|} \tag{3}$$

is adiabatically conserved, the particles are always trapped on the same $|B|$ position. Therefore, if energetic ions are trapped in the closed $|B|$ contour region, they can be confined. It means that the loss conditions with respect to the trapped position are altered by the diamagnetic effect. The color depth representing the tracing time of the alpha particles is shown on the initial position of them in Fig. 4. These particles were started from $\varphi = \pi/2$ with zero parallel velocity. These figures clearly show that there are two opposing effects from the diamagnetic effect: reduction of loss particles which are trapped in the closed $|B|$ contour region and increase of them which are trapped in further inner core region. Then, we have to clarify which effect is more effective.

The black dots in Fig. 4 show the first trapped points of 10,000 fusion alpha particles, where passing particles were ignored. For these birth profiles, energetic ion orbits were calculated using the F3D-OFMC code [6]. Figure 5 shows the loss rate which is defined by the ratio between the total initial energy and the total

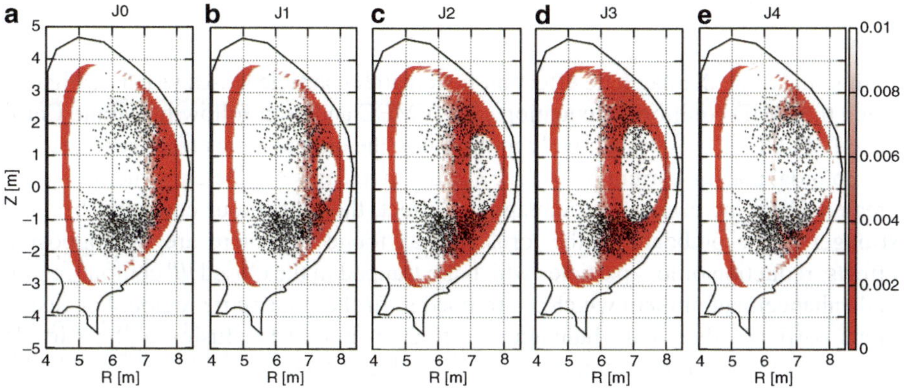

Fig. 4 The color depth shows the tracing time [s] for alpha particles on the initial (R,Z) position. These particles started from $\varphi = \pi/18$ with no parallel velocity on the initial position. The black dots show the first reflect point of 10,000 fusion alpha particles (i.e., passing particles were ignored in these figures)

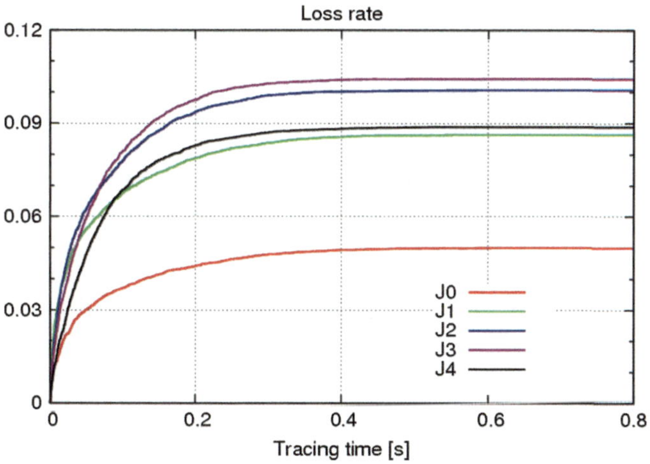

Fig. 5 The tracing time dependence of the loss rate for each field model, J0, J1, J2, J3 and J4

energy of the loss particles which collided with the first wall. The final loss rate for J0, J1, J2, J3 and J4 models were 0.050, 0.086, 0.101, 0.104 and 0.089, respectively. It was found that the weak diamagnetic effect increased the fusion alpha particle losses, while the strong diamagnetic effect reduced them. If we can choose the conditions for the strong diamagnetic system, such as the high pressure, the high aspect ratio and the high safety factor, the energetic ions can be well confined. In a future work, we will investigate the diamagnetic effect on the energetic ion losses by changing the aspect ratio and the safety factor.

4 Summary

In this study, the diamagnetic effect on fusion alpha particle losses was investigated using the MHD equilibrium equation code VMEC and orbit following Monte Carlo code F3D-OFMC. By respectively multiplying the diamagnetic effect generated by the poloidal plasma current by 0, 1, 2, 3 and 4, five cases of 3D field structures were created. With increasing the diamagnetic effect, the $|B|$ contour curvature become stronger at the outboard of the torus. It was found that there are two opposing effects: reduction of loss particles which are trapped in the closed $|B|$ contour region and enhancement of them which are trapped in further inner core region. For fusion alpha particles, the weak diamagnetic effect increased the fusion alpha particle losses, while the strong diamagnetic effect reduced them.

Acknowledgment The authors were grateful for the support of the Ministry of Education, Culture, Sports, Science and Technology of Japan via "Energy Science in the Age of Global Warming" of Global Center of Excellence (G-COE) program (J-051).

References

1. Shinohara K et al (2011) Nucl Fusion 51:063028
2. Strumberger E et al (2010) Nucl Fusion 50
3. Suzuki Y et al (2003) Nucl Fusion 43:406415
4. Hirshman SP, Meier HK (1985) Phys Fluids 28:1387
5. Hirshman SP, Merkel WIRP (1986) Comput Phys Commun 43:143
6. Tani K et al (1981) J Phys Soc Jpn 50:1726

Formulation of the Two-Dimensional Heat Transport Equation in Tokamak Plasmas

Haruki Seto and Atsushi Fukuyama

Abstract It is desirable to self-consistently describe two-dimensional transport in a whole tokamak plasma including both core and peripheral regions, since the pedestal region lying between the core and peripheral regions strongly affects the confinement property of the core plasma. The set of two-dimensional transport equations with poloidal angle dependence has been formulated from Braginskii's multi-fluid equations and the neoclassical transport theory and has been coupled with electromagnetic equations in order to develop a two-dimensional transport simulation code TASK/T2. The set of transport equations consists of equations of the particle density and the momentum (radial, parallel, and toroidal), and the heat transport equation, which provide the consistent description of transport phenomena through the whole region. In this paper, we focus on the derivation of the heat transport equation.

Keywords Multi-fluid model • Tokamak • Two-dimensional transport model

1 Introduction

In most of conventional transport simulations in tokamaks, transport in a core region is usually described as a one-dimensional problem by flux-surface average. On the other hand, transport in a peripheral SOL-divertor region is usually described as a two-dimensional problem with simplified transport models and plasma flow. Since the pedestal region strongly affects the confinement property of core plasmas, however, it is desirable to self-consistently describe two-dimensional transport in a whole plasma including both core and peripheral regions. Recent remarkable progress in computational resources has made such analyses feasible. In the present study, a set of two-dimensional transport equations [1, 2] has

H. Seto (✉) • A. Fukuyama
Kyoto University, Kyoto, Japan
e-mail: seto@p-grp.nucleng.kyoto-u.ac.jp; fukuyama@nucleng.kyoto-u.ac.jp

T. Yao (ed.), *Zero-Carbon Energy Kyoto 2012*, Green Energy and Technology, DOI 10.1007/978-4-431-54264-3_24,

been formulated with the neoclassical viscosity tensor [3] in a magnetic flux coordinate system from must-fluid equations. Transport equations consist of the equation for particle density, momentum (radial, parallel, toroidal), and energy transport for both electrons and ions. The set of transport equations is coupled with a set of equations for electromagnetic field in order to develop an two-dimensional transport simulation code TASK/T2. The set of electromagnetic equations consists of Grad–Shaftranov equation, magnetic diffusion equation [4], and Poisson equation for the static electric field. This model includes the effects of the neoclassical viscous force and the field evolution and will make it possible to describe the time evolution of pedestal structure and the interaction between the core and the peripheral plasma self-consistently. In this paper, we focus on the two-dimensional formulation of the heat transport equations for electrons and ions. Coordinate systems and assumptions are described in Sect. 2. In Sect. 3, the two-dimensional heat transport equation is derived from multi-fluid equations including the neoclassical viscosity [3]. Finally a short summary is given in Sect. 4.

2 Coordinate Systems and Assumptions

We employ magnetic flux coordinate system (MFCS), $(\xi_1^M, \xi_2^M, \xi_3^M) = (\rho, \chi, \zeta)$ and a vector $v_i^{\mathrm{M}} \equiv v \cdot \nabla \xi_i^M$, to express spatial variations of quantities in MHD equilibrium and neoclassical transport coordinate system (NTCS), $(e_1^{\mathrm{N}}, e_2^{\mathrm{N}}, e_3^{\mathrm{N}}) = (\nabla\rho, B/B, R^2\nabla\zeta)$ and $v_i^{\mathrm{N}} \equiv v \cdot e_i^{\mathrm{N}}$, to express components of vector quantities for the compatibility with the neoclassical transport theory [3] as shown in Fig. 1. In this study the following two assumptions are made: (1) toroidal axisymmetry and (2) phenomena at the Alfvén velocity are much faster than diffusion of the magnetic field and transport phenomena.

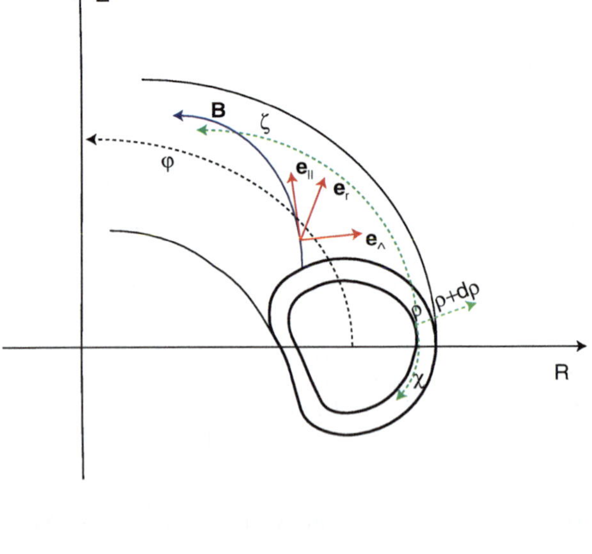

Fig. 1 Coordinate systems

3 Formulation of the Two-Dimensional Heat Transport Equation

We derive the two-dimensional transport model from Braginskii's multi-fluid equations [5] which consist of equation of continuity, equation of motion and equation of energy transport.

$$\frac{\partial n_a}{\partial t} + \nabla \cdot (n_a u_a) = S_{na} \tag{1}$$

$$\frac{\partial}{\partial t}(m_a n_a u_a) + \nabla \cdot (m_a n_a u_a u_a) = -\nabla p_a - \nabla \cdot \overleftrightarrow{\pi}_a + e_a n_a (E + u_a \times B) + F_a + S_{ma} \tag{2}$$

$$\frac{\partial}{\partial t}\left(\frac{3}{2}p_a\right) + \nabla \cdot \left(q_a + \frac{5}{2}p_a u_a + \overleftrightarrow{\pi}_a \cdot u_a\right) = u_a \cdot \left(\nabla \cdot \overleftrightarrow{\pi}_a\right) + u_a \cdot \nabla p_a + Q_a + S_{E_a} - \frac{1}{2} m_a u_a^2 S_{na} \tag{3}$$

where a denotes particle species, n_a is the density of particle, u_a is the flow velocity, S_{n_a} is the source of particle, m_a is the mass, p_a is the isotropic pressure, $\overleftrightarrow{\pi}_a$ is the anisotropic pressure tensor, e_a is the electric charge, E is the electric field, B is the magnetic field, F_a is the friction force, S_{ma} is the source of momentum, q_a is the heat flux, Q_a is the energy exchange term, and S_{Ea} is the source of energy. In this study, we employ the neoclassical viscosity tensor [3] as the anisotropic pressure tensor

$$\overleftrightarrow{\pi}_a = \pi_a^{\text{neo}}\left(e_2^{\text{N}} e_2^{\text{N}} - \frac{1}{3}\overleftrightarrow{I}\right), \quad \pi_a^{\text{neo}} = p_{a\parallel} - p_{a\perp} \tag{4}$$

where π_a^{neo} is the neoclassical parallel viscosity coefficient. The transport model consists of the equation of particle density, momentum and heat transport for electrons and ions. The detail of the derivation of the equations of particle density and momentum is given in [1, 2] and we focus on the derivation of the heat transport equation. The LHS of Eq. (3) consists of the time derivative term and the divergence terms of the heat flux q_a, the internal energy flow $\frac{5}{2}p_a u_a$, and the anisotropic pressure tensor $\overleftrightarrow{\pi}_a \cdot u_a$ and the RHS of Eq. (3) consists of the viscous heating term $u_a \cdot \left(\nabla \cdot \overleftrightarrow{\pi}_a\right)$, the advective pressure gradient term $u_a \cdot \nabla p_a$, the collisional energy exchange term Q_a, and the energy source term $S_{Ea} - \frac{1}{2} m_a u_a^2 S_{na}$. We employ the Braginskii's heat flux [5] as q_a, which can be written as the following form.

$$q_a = -n_a \chi_{a\parallel} \nabla_{\parallel} T_a - n_a \chi_{a\wedge} \nabla_{\wedge} T_a - n_a \chi_{a\perp} \nabla_{\perp} T_a + p_a V^{hp} \tag{5}$$

where $\chi_{a\,\parallel}$, $\chi_{a\,\wedge}$, and $\chi_{a\,\perp}$ are Braginskii's thermal diffusion coefficients in the parallel, cross, and perpendicular directions, respectively, and V^{hp} is the heat pinch velocity. In addition, we assume the heat flux can be written as the following form.

$$q_a = -n_a \overleftrightarrow{\chi}_a \cdot \nabla T_a + p_a V^{hp} \tag{6}$$

Taking the scalar product of Eqs. (5) and (6) with $\nabla\xi_i{}^M$, we obtain following the relation.

$$\begin{aligned} q_a \cdot \nabla\xi_i^M =& - n_a\chi_{a\parallel}\left(\nabla\xi_i^M \cdot e_2^{\mathrm{N}}\right)\left(e_2^{\mathrm{N}} \cdot \nabla T_a\right) - n_a\chi_{a\wedge}\nabla\xi_i^M \cdot \left(e_2^{\mathrm{N}} \times \nabla T_a\right) \\ & - n_a\chi_{a\perp}\left[\nabla\xi_i^M - \left(\nabla\xi_i^M \cdot e_2^{\mathrm{N}}\right)e_2^{\mathrm{N}}\right] \cdot \nabla T_a + p_a \nabla\xi_i^M \cdot V_a^{\mathrm{hp}} \\ =& - n_a\chi_{a\parallel}\frac{\psi'}{\sqrt{g}B^2}\left(\frac{\psi'}{\sqrt{g}}\epsilon_{i31} + I g^{i3}\right)\frac{\partial T_a}{\partial\xi_2^M} \\ & - n_a\chi_{a\wedge}\frac{1}{B}\left(\psi' g^{1i}g^{3j} - \psi' g^{3i}g^{1j} + \frac{I}{\sqrt{g}}\epsilon_{ji3}\right)\frac{\partial T_a}{\partial\xi_j^M} \\ & - n_a\chi_{a\perp}g^{ij}\frac{\partial T_a}{\partial\xi_j^M} + n_a\chi_{a\perp}\frac{\psi'}{\sqrt{g}B^2}\left(\frac{\psi'}{\sqrt{g}}\epsilon_{i31} + I g^{i3}\right)\frac{\partial T_a}{\partial\xi_2^M} + p_a V_i^{\mathrm{hp,M}} \end{aligned} \tag{7}$$

$$q_a \cdot \nabla\xi_i^M = -\sum_{j=1}^{3} n_a \chi_{aij}^{\mathrm{M}} \frac{\partial T_a}{\partial \xi_j^M} + p_a V_i^{\mathrm{hp,M}} \tag{8}$$

where g^{ij} is the contravariant metric tensor in MFCS and ε_{ijk} is the Levi–Civita symbol. By using the relation $\chi_{a\,\parallel} \gg \chi_{a\,\wedge} \gg \chi_{a\,\perp}$ and ignoring the heat pinch term for simplicity, the divergence of the heat flux in MFCS can be written as follows.

$$\nabla \cdot q_a = -\sum_{i=1}^{3}\sum_{j=1}^{3} \frac{1}{\sqrt{g}} \frac{\partial}{\partial \xi_i^M}\left(\sqrt{g} n_a \chi_{aij}^{\mathrm{M}} \frac{\partial T_a}{\partial \xi_j^M}\right) \tag{9}$$

$$\chi_{aij}^{\mathrm{M}} = \begin{pmatrix} g^{11}\chi_{a\perp} & -\dfrac{I}{\sqrt{g}B}\chi_{a\wedge} & \dfrac{\psi' g^{11}}{BR^2}\chi_{a\wedge} \\ \dfrac{I}{\sqrt{g}B}\chi_{a\wedge} & \left(\dfrac{\psi'}{\sqrt{g}B}\right)^2\chi_{a\parallel} & \dfrac{\psi' g^{12}}{BR^2}\chi_{a\wedge} \\ -\dfrac{\psi' g^{11}}{BR^2}\chi_{a\wedge} & \dfrac{\psi' I}{\sqrt{g}B^2R^2}\chi_{a\parallel} & \frac{1}{R^2}\chi_{a\perp} \end{pmatrix} \tag{10}$$

By using the transform matrix from NTCS to MFCS $\mathcal{T}_{ij}^{MN}$, we obtain the divergence of the energy flow in MFCS as follows:

$$\nabla\cdot\left(\frac{5}{2}p_a u_a\right)=\sum_{i=1}^{3}\sum_{j=1}^{3}\frac{1}{\sqrt{g}}\frac{\partial}{\partial\xi_i^M}\left(\sqrt{g}\frac{5}{2}\mathcal{T}_{ij}^{MN}p_a u_{aj}^{N}\right) \tag{11}$$

Since the neoclassical viscosity tensor is in the CGL form, the divergence of the anisotropic pressure tensor can be written as the following simple form:

$$\begin{aligned}\nabla\cdot\left(\overleftrightarrow{\pi}_a\cdot u_a\right)&=\nabla\cdot\left[\pi_a^{\mathrm{neo}}\left(e_2^{\mathrm{N}}e_2^{\mathrm{N}}-\frac{1}{3}\overleftrightarrow{I}\right)\cdot u_a\right]\\&=\frac{1}{\sqrt{g}}\frac{\partial}{\partial\xi_2^M}\left(\frac{\pi_a^{\mathrm{neo}}\psi'}{B}u_2^{\mathrm{N}}\right)-\sum_{i=1}^{3}\sum_{j=1}^{3}\frac{1}{\sqrt{g}}\frac{\partial}{\partial\xi_i^M}\left(\sqrt{g}\frac{\pi_a^{\mathrm{neo}}\mathcal{T}_{ij}^{MN}}{3}u_{aj}^{\mathrm{N}}\right)\end{aligned} \tag{12}$$

As the divergence of the neoclassical viscosity tensor $\nabla\cdot\overleftrightarrow{\pi}_a$ can be written as

$$\nabla\cdot\overleftrightarrow{\pi}_a=\pi_a^{\mathrm{neo}}\left(\kappa-e_2^{\mathrm{N}}e_2^{\mathrm{N}}\cdot\nabla\ln B\right)+e_2^{\mathrm{N}}e_2^{\mathrm{N}}\cdot\nabla\pi_a^{\mathrm{neo}}-\frac{1}{3}\nabla\pi_a^{\mathrm{neo}}, \tag{13}$$

we obtain the viscous heating term as follows.

$$\begin{aligned}u_a\cdot\left(\nabla\cdot\overleftrightarrow{\pi}_a\right)=&\sum_{i=1}^{3}\sum_{j=1}^{3}\mathcal{T}_{ij}^{MN}\left(\sum_{k=1}^{3}\pi_a^{\mathrm{neo}}g_{ik}\kappa_k^{\mathrm{M}}-\frac{1}{3}\frac{\partial\pi_a^{\mathrm{neo}}}{\partial\xi_i^M}\right)u_{aj}^{\mathrm{N}}\\&+\frac{\psi'}{\sqrt{g}B}\left(\frac{\partial\pi_a^{\mathrm{neo}}}{\partial\xi_2^M}-\pi_a^{\mathrm{neo}}\frac{\partial\ln B}{\partial\xi_2^M}\right)u_{a2}^{\mathrm{N}}\end{aligned} \tag{14}$$

where κ is the curvature vector of the magnetic field and g_{ij} is the covariant metric tensor in MFCS. Finally the advective pressure gradient term and the collisional energy exchange term can be written in MFCS as follows.

$$u_a\cdot\nabla p_a=\sum_{i=1}^{3}\sum_{j=1}^{3}\mathcal{T}_{ij}^{MN}u_{aj}^{\mathrm{N}}\frac{\partial p_a}{\partial\xi_i^M} \tag{15}$$

$$Q_{ab}=\frac{3}{2}n_a\frac{T_b-T_a}{\tau_{ab}},\quad \tau_{ab}=\frac{3\pi\sqrt{2\pi}\epsilon_0^2 m_a m_b}{n_b e^4 Z_a^2 Z_b^2\ln\Lambda_{ab}}\left(\frac{T_a}{m_a}+\frac{T_b}{m_b}\right)^{3/2} \tag{16}$$

Substituting Eqs. (9), (11), (12), (14)–(16) into Eq. (3), we obtain the heat transport equation in MFCS as follows.

$$
\begin{aligned}
&\frac{\partial}{\partial t}\left(\frac{3}{2}p_a\right)-\sum_{i=1}^{3}\sum_{j=1}^{3}\frac{1}{\sqrt{g}}\frac{\partial}{\partial \xi_i^{M}}\left(\sqrt{g}n_a\chi_{aij}^{\mathrm{M}}\frac{\partial T_a}{\partial \xi_j^{M}}\right)+\sum_{i=1}^{3}\sum_{j=1}^{3}\frac{1}{\sqrt{g}}\frac{\partial}{\partial \xi_i^{M}}\left(\sqrt{g}\frac{5}{2}\mathcal{T}_{ij}^{MN}u_{aj}^{\mathrm{N}}p_a\right)\\
&+\frac{1}{\sqrt{g}}\frac{\partial}{\partial \xi_2^{M}}\left(\frac{\pi_a^{\mathrm{neo}}\psi'}{B}u_{a2}^{\mathrm{N}}\right)-\sum_{i=1}^{3}\sum_{j=1}^{3}\frac{1}{\sqrt{g}}\frac{\partial}{\partial \xi_i^{M}}\left(\sqrt{g}\frac{1}{3}\pi_a^{\mathrm{neo}}\mathcal{T}_{ij}^{MN}u_{aj}^{\mathrm{N}}\right)\\
&=\sum_{i=1}^{3}\sum_{j=1}^{3}\mathcal{T}_{ij}^{MN}\left(\sum_{k=1}^{3}\pi_a^{\mathrm{neo}}g_{ik}\kappa_k^{\mathrm{M}}-\frac{1}{3}\frac{\partial \pi_a^{\mathrm{neo}}}{\partial \xi_i^{M}}\right)u_{aj}^{\mathrm{N}}+\frac{\psi'}{\sqrt{g}B}\left(\frac{\partial \pi_a^{\mathrm{neo}}}{\partial \xi_2^{M}}-\pi_a^{\mathrm{neo}}\frac{\partial \ln B}{\partial \xi_2^{M}}\right)u_{a2}^{\mathrm{N}}\\
&+\sum_{i=1}^{3}\sum_{j=1}^{3}\mathcal{T}_{ij}^{MN}u_{aj}^{\mathrm{N}}\frac{\partial p_a}{\partial \xi_i^{M}}+\sum_{b\neq a}\frac{3}{2}n_a\frac{T_b-T_a}{\tau_{ab}}+S_{Ea}-\frac{1}{2}m_a u_a^2 S_{na}
\end{aligned}
\tag{17}
$$

To close our heat transport equation, appropriate two-dimensional modeling on the neoclassical viscosity coefficient is needed. We employ the neoclassical viscosity coefficient for multi-collisionality regime [6] in this formulation. Our two-dimensional heat transport equation including the neoclassical viscosity tensor is formulated in a general axisymmetrical MFCS without simplified plasma flow assumptions, which is applicable not only in a high collisionality (low temperature) regime but also on the low collisonality (high temperature) regime. It is an advantage in the case of two-dimensional transport analyses in the whole tokamak plasma including both high and low collisionality regime, since conventional two-dimensional models used in a peripheral region are formulated with simplified plasma flow assumptions and high collisonality closures.

4 Summary

We have formulated the heat transport equation from a set of multi-fluid equations with the neoclassical viscosity tensor [3, 6] and the Braginskii's heat flux [5] in MFCS, which provides the consistent description of transport phenomena throughout the tokamak plasma. It can be reduced to two-dimensional owing to toroidal axisymmetry. By combining the heat transport equation with other transport equations (1), (2) and the set of electromagnetic equations which consist of the equation for static electric field, the magnetic diffusion equation and Grad-Shafranov equation, a self-consistent two-dimensional transport analysis including the field evolution will be available.

Acknowledgements This work is supported by Grant-in-Aid for Science Research (S) (20226017) from JSPS and G-COE program (J-051) from MEXT.

References

1. Seto H, Fukuyama A (2012) Zero-Carbon Energy Kyoto 2011. Green Energy and Technology, pp 311–317
2. Seto H, Fukuyama A (2012) Contrib Plasma Phys 52(5–6):384–390
3. Hirshman SP, Sigmar DJ (1981) Nucl Fusion 21(9):1079
4. Hinton FL, Hazeltine RD (1976) Rev Mod Phys 48:239
5. Braginskii SI (1965) In: Leontovich MA (ed) Reviews of plasma physics, vol 1. Consultants Bureau, New York, p 205
6. Callen JD, UW-CPTC 09-6R (2010). Report available from http://www.cptc.wisc.edu/~callen/UW-CPTC_09-6R.pdf

Application of Very High Speed Camera in Measurement of Liquid Film Flow on Nuclear Rod Bundle in Micro-Scale

Son Hong Pham, Zensaku Kawara, and Tomoaki Kunugi

Abstract The annular liquid film flow on nuclear rod bundle is the complicated phenomena with very fine scales on time and space, so existing measuring techniques have many difficulties to investigate this phenomenon. In this work, to avoid these problems, a very high speed camera with long-ranged microscope optical system has been introduced to measure the flow on the simulated 3 × 3 rod bundle made of transparent acrylic resin. The results include the real-time visualization of important processes of entrainment and deposition with high spatiotemporal resolution and the relationship between velocity and diameter of each droplet. These initial data have approved the advantage of high speed camera technique in the studying of annular two phase flow.

Keywords Annular liquid film • High speed camera • Nuclear rod bundle • Two-phase flow

1 Introduction

Concerning directly to the safety of boiling water reactor, the liquid film flow on nuclear fuel rods has become the subject of many studies. This annular two-phase flow is characterized by a number of factors including the thin liquid layers distributed on the rods' surface, the core of gas flowing through the space between the rods at high velocity, and liquid droplets generated or vanished due to the processes of entrainment and deposition, respectively. Because the effectiveness of the heat transfer process from fuel rods strongly depends on the liquid film, many techniques of measurement have been applied to study the flow such as ultrasonic

S.H. Pham • Z. Kawara (✉) • T. Kunugi
Department of Nuclear Engineering, Kyoto University, Yoshida, Sakyo, Kyoto 606-8501, Japan
e-mail: kawara@nucleng.kyoto-u.ac.jp

T. Yao (ed.), *Zero-Carbon Energy Kyoto 2012*, Green Energy and Technology,
DOI 10.1007/978-4-431-54264-3_25,

transmission, conductivity probe, laser induced fluorescence, etc. The obtained experimental data, however, are still not enough to construct the physical model for critical heat flux analysis especially at the micro-scale. Remaining problems are mainly caused by the complicated geometry of fuel rod bundles, extra high velocity and unstable interface of liquid and gas flow.

Previously, the annular two-phase flow has been widely studied by electrical conductivity method based on the relationship of the thickness and conductance of the liquid film. Feldhaus et al. [1] used five pairs of conductance probes to determine the average film thickness for a duct simulating 5×2 rod bundle. The data was recorded at five points at a sampling frequency of 1 kHz and showed the effect of different types of spacer on the flow. More recently, Damsohn et al. [2] introduced a new sensor consisting of electrodes flush to the rod surface. The sensor could measure the film thickness and locate droplets deposition at 64×16 point with a spatial resolution of 3.12 mm^2 and time resolution of 10 kHz.

Another method of film thickness measurement is ultrasonic. Kamei [3] applied a sensor with rotating reflector installed inside a simulated fuel rod. The sensor could measure time-dependent spatial distribution of liquid film thickness around the rod and needed 4 ms for reconstruction of one image of circumferential. Although being considered as a non-intrusive method, this technique is limited method for measuring small film thickness as well as the determination of small change in wave surfaces because of poor spatiotemporal resolution caused by sound wavelength.

Not only give the data in signal form like above methods, laser induced fluorescence technique provides visual information of the liquid film. Most recently, Schubring et al. [4, 5] performed planar laser-induced fluorescence measurements. A small concentration of fluorescent tracer was introduced into liquid phase, causing the liquid film to appear as bright regions when exposed to laser light. The images recorded by a camera give the visual information of gas–liquid interface. Four hundred images recorded by a camera for each flow condition at time step of 60 ms provided the visual information of gas–liquid interface with the estimated uncertainty of 25 μm. However, at this time step, the data cannot describe the fast processes like entrainment and deposition.

To get over these difficulties and archive the capability of studying the flow at micro-scale, a new approach using a very high speed digital camera system (more than 10,000 fps) and a Cassegrain optical lens have been introduced in this work. A simulated 3×3 rectangular rod bundle made of acrylic resin allows the full visual observation of this system. The results include real-time visualization of important processes such as entrainment and deposition, and relationship between velocity and diameter of droplets. The data obtained at spatiotemporal resolution up to 7 μm and 20 μs have approved the effectiveness of high speed camera technique.

2 Experiment

2.1 Two-Phase Flow Loop

Figure 1 shows the experimental facility. From water tank, the water is driven into mixing section with porous metal tubes. Water goes to the outer surface of tubes through porous parts and is pushed by the air coming from gas compressor to form upward liquid film flow. Three precise spacers assign the tubes into 3 × 3 rectangular grid. After leaving the test section, two phase flow goes into the separator. Then the gas is released to environment while the water comes back to the water tank.

The valves on the water and gas inlet pipes were adjusted to maintain the average speed of gas and water in the test section at 25 and 0.11 m/s, respectively. A strong effort has been put on the support frame of the test section to minimize the vibration caused by two-phase flow.

The information about the spacers and the cross-section of test section is presented in Fig. 2. The minimum distance between two simulated rods is 3 mm.

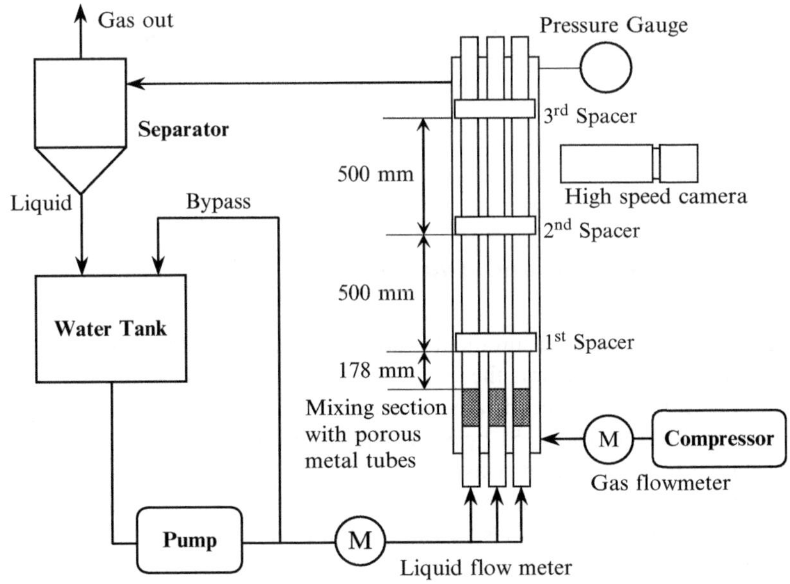

Fig. 1 Schematic of experimental facility

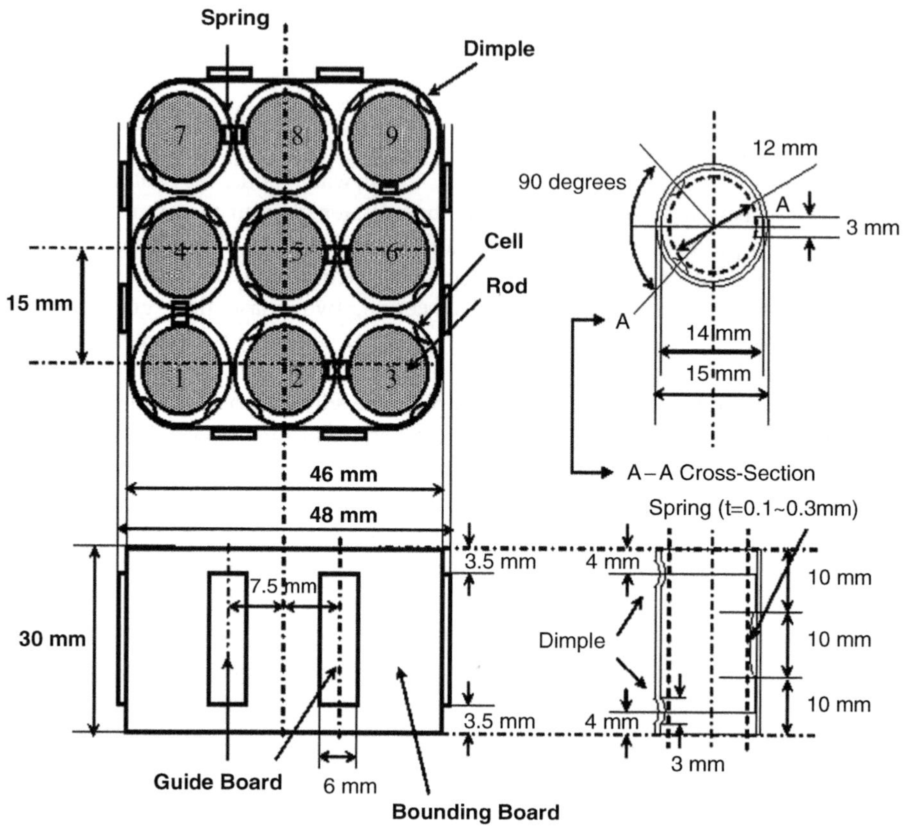

Fig. 2 Spacer and the cross-section of test section

2.2 Data Acquisition and Analysis

In the present work, the measurement point was located at 100 mm above the 2nd spacer. The focus position was the space between rod number 4 and rod number 7 shown in Fig. 2. The high speed camera recorded image data at the rate of 8,200 fps with the resolution of 512×512 pixels. The length of each movie clip was about 0.4 s and total number of clips was 50.

Then image processing was done by using Cine Viewer developed by Vision Research Company to obtain a visualization of important processes and quantitative information of droplets in the gas core.

Using the minimum distance between these two rods of 3 mm, an image was taken without water flowing through the test section to perform the calibration for image data. The calibration showed that one image pixel is equivalent to $7 \times 7\ \mu m^2$.

3 Results and Discussion

3.1 Visualization of Important Processes of Entrainment and Deposition

The processes of entrainment and deposition are presented in Figs. 3 and 4, respectively. The time step between successive images is 72 μs in these two series.

The details of image show the capability of analyzing many characteristics of liquid film such as thickness, shape of wave surfaces, and the mechanism of important processes of entrainment and deposition.

3.2 Droplets' Velocity and Diameter

Figure 5 shows the relationship between velocity and diameter of droplets. Totally, 100 droplets flying through the measurement area were taken into account.

The data show the common trend of the relationship in which droplet velocity decreases with the increase in diameter. There are some small droplets moving at low velocities because they are created near the measuring location and have not obtained enough acceleration from gas flow.

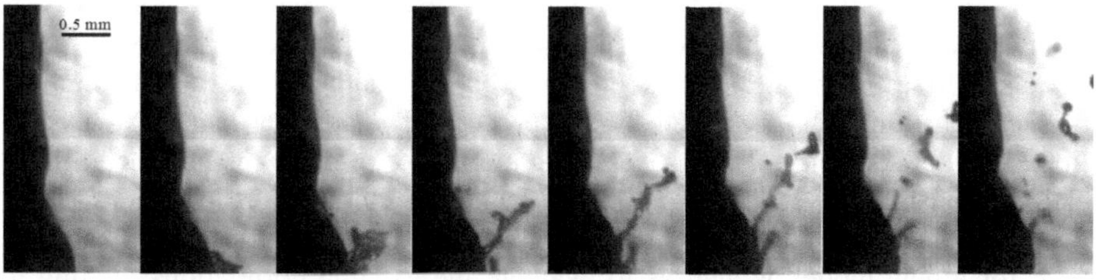

Fig. 3 The process of entrainment—droplets are created from the surface of liquid film

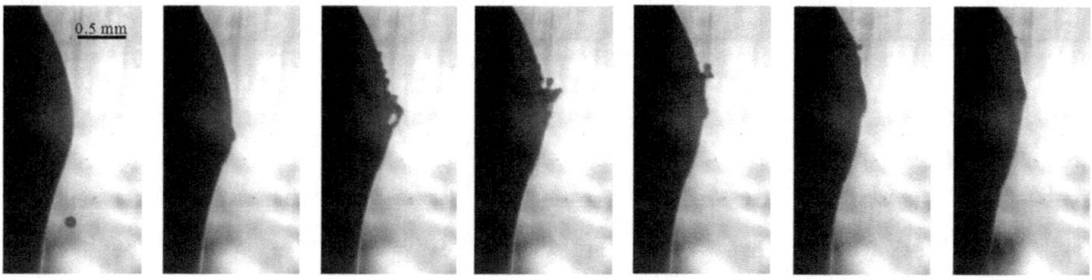

Fig. 4 The process of deposition—a droplet goes into the surface of liquid film

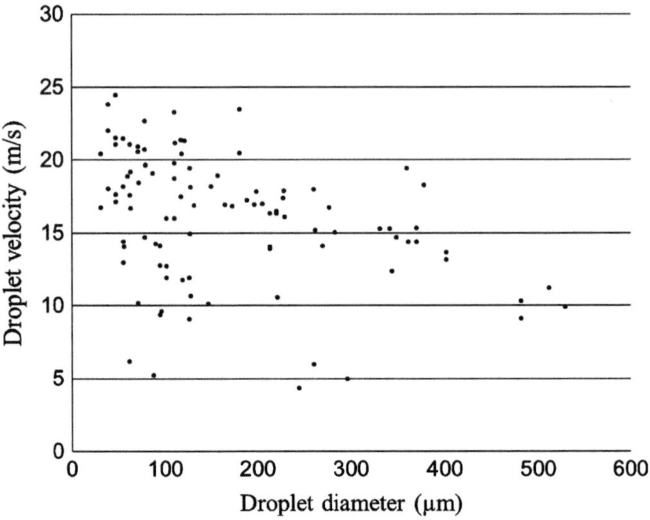

Fig. 5 Relationship between velocity and diameter of each droplet

4 Conclusion

The direct observations of important processes in very high spatiotemporal resolutions as well as the ability of quantitative measurement are the advantages of high speed camera technique in the studying of annular two phase flow.

Using these capabilities, the experiment is being continued to study the number of secondary droplets generated by the process of deposition and the length of liquid "tail" occurring in the process of entrainment in different flow conditions. The relationship between the liquid film thickness and flow rate, the space distribution of droplets, etc. will be also determined to build the databases which could be used for a new model as well as to evaluate the existing correlations relating to upward annular liquid film flow.

Acknowledgment This work was partly supported by an "Energy Science in the Age of Global Warming" of Global Center of Excellence (G-COE) program (J-051) of the Ministry of Education, Culture, Sports, Science and Technology of Japan.

References

1. Feldhaus G, Azzopardi BJ, Zeggel W (2002) Annular flow experiments in rod bundles with spacers. Nucl Eng Des 213:199–207
2. Damsohn M, Prasser HM (2009) High-speed liquid film sensor for two-phase flows with high spatial resolution based on electrical conductance. Flow Meas Instrum 20:1–14

3. Kamei T, Serizawa A (1998) Measurement of 2-dimensional local instantaneous liquid film thickness around simulated nuclear fuel rod by ultrasonic transmission technique. Nucl Eng Des 184:349–362
4. Schubring D, Ashwood AC, Shedd TA, Hurlburt ET (2010) Planar laser-induced fluorescence (PLIF) measurements of liquid film thickness in annular flow. Part I: Methods and data. Int J Multiphase Flow 36:815–824
5. Schubring D, Ashwood AC, Shedd TA, Hurlburt ET (2010) Planar laser-induced fluorescence (PLIF) measurements of liquid film thickness in annular flow. Part II. Analysis comparison models. Int J Multiphase Flow 36:825–835

Evaluation of Tritium Production in LiPb Blanket System Using Neutrons Analysis

Saerom Kwon, Kazuyuki Noborio, Ryuta Kasada, and Satoshi Konishi

Abstract Tritium production in LiPb blanket was evaluated by neutronics analysis with Monte Carlo code, MCNP using nuclear cross-section data from the FENDL-2.1 libraries. The result suggests the sufficient TBR (Tritium Breeding Ratio) can be obtained without using additional neutron multiplier such as beryllium. Based on the calculation, a plan for the integral experiment of a small LiPb system to verify it with our DD fusion neutron beam source was made. An integral experiment on LiPb with DD neutrons properly, transport and absorption of the neutron in a small test module was evaluated.

Keywords Fusion reactor blanket • Integral experiment • LiPb • MCNP code • Tritium production

1 Introduction

Fusion is regarded as one of the viable energy resources that contribute to the global climate change problem as it has minimal carbon dioxide emission and significant potential capability to supply large amount of energy with virtually no resource constraints. For the utilization of a nuclear fusion reactor, tritium fuel must be self-supplied by the production/breeding with the lithium containing blanket.

This study intends to numerically evaluate the tritium production in the modules consist of same materials of blankets to be used in the reactor for the comparison with the examination by neutron sources to verify the actual performance. ITER TBM (Test Blanket Module) and other integral experiment will benchmark

S. Kwon (✉)
Graduate School of Energy Science, Kyoto University, Gokasho, Uji, Kyoto 611-0011, Japan
e-mail: saerom.kwon@iae.kyoto-u.ac.jp

K. Noborio • R. Kasada • S. Konishi
Institute of Advanced Energy, Kyoto University, Gokasho, Uji, Kyoto 611-0011, Japan

T. Yao (ed.), *Zero-Carbon Energy Kyoto 2012*, Green Energy and Technology, 239
DOI 10.1007/978-4-431-54264-3_26, © Springer Japan 2013

the neutronics calculations so that 3D calculation for the entire reactor would be confirmed. We carried out neutronics analysis on the tritium production behavior in the LiPb–SiC blanket module for the preparation of the integral experiment and optimization of its experiment conditions such as the size of LiPb, the enrichment rate of lithium-6.

2 Calculation Method and Models

The tritium production in LiPb blanket module was calculated with the MCNP code [1] and nuclear data library: FENDL-2.1 [2]. We calculated two models for the purposes of (1) determination of the size of suitable LiPb block for integral experiment with small scale DD neutron beam source we plan to prepare in Kyoto University, (2) determination of the target enrichment of lithium-6 in LiPb for the sufficient tritium production. All of the calculations used the rectangular coordinate and their schematics are shown in Fig. 1. Neutrons are assumed to be generated in two kinds of monochromic energy, 2.45 MeV (from DD reaction) for the experiment plan and 14.1 MeV (from DT reaction) from a point source for comparison with the results of DD neutrons. The distance of the neutron source set 10 cm from the target surface, it is an assumed actual experiment condition that position of neutron source.

2.1 *Calculation Model 1*

In calculation model 1, effects of the size of LiPb block on the tritium production were evaluated. Considering the coverage of the entire surface of the burning plasma in the torus would be limited to around 0.7, in order to obtain the net tritium self-sufficiency, at least 1.43 of the TBR will be required with 14.1 MeV neutrons.

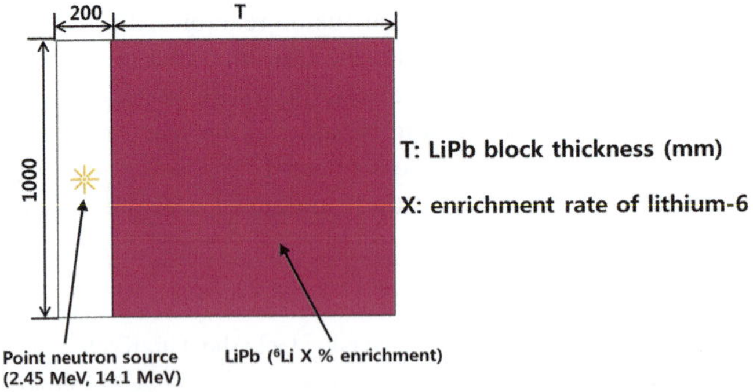

Fig. 1 Image of the calculation model

Calculation No.	Side length (cm)	Volume (cm^3)
1	20	8000
2	40	64000
3	60	216000
4	80	512000
5	100	1000000

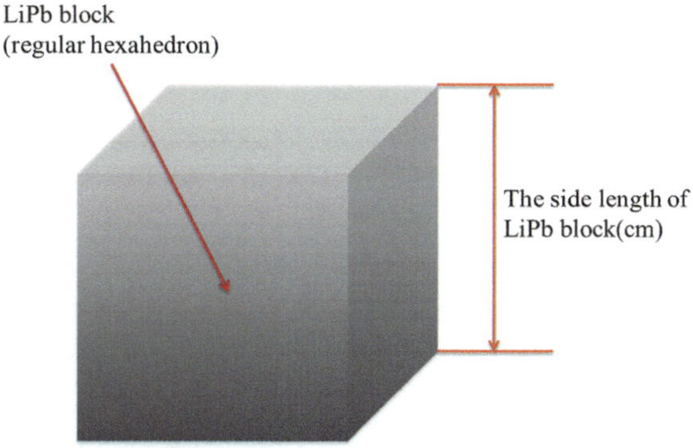

Fig. 2 The size table and the schematic of calculation model 2

Calculated sizes of the LiPb block was regular hexahedron in shape and the lengths of each side were 20 cm, 40 cm, 60 cm, 80 cm and 100 cm with 14.1 MeV, 2.45 MeV respectively. The calculation model 1 and its size information are shown in Fig. 2.

2.2 Calculation Model 2

The composition of LiPb considered as a breeding material of the blanket is Li: Pb = 17:83. Since the natural Lithium contains lithium-6 that produces tritium with thermal neutron as minority (^{6}Li:^{7}Li = 7.5:92.5), it is necessary to enrich the lithium-6. In the calculation model 2 shown in Fig. 3, enrichment of lithium-6 was changed from 0%(only lithium-7 in the Lithium contents of the LiPb) to 100%. Its module size is chosen from the results of the calculation with model 1.

3 Results of Calculation and Discussion

3.1 Calculation Model 1

TBR with various LiPb (lithium-6 90% enrichment) size is shown in the Fig. 4. From a viewpoint of benchmarking TBR around 1.43, the required size of LiPb module is between 40 cm and 60 cm with 14.1 MeV neutron source. The observed difference between 14.1 MeV and 2.45 MeV, is considered that the while lead in LiPb works as neutrons multiplier, however it does not occur (n, 2n) reaction of lead at 2.45 MeV neutrons, and thus tritium production of lithium-6 with thermal neutron showed different value.

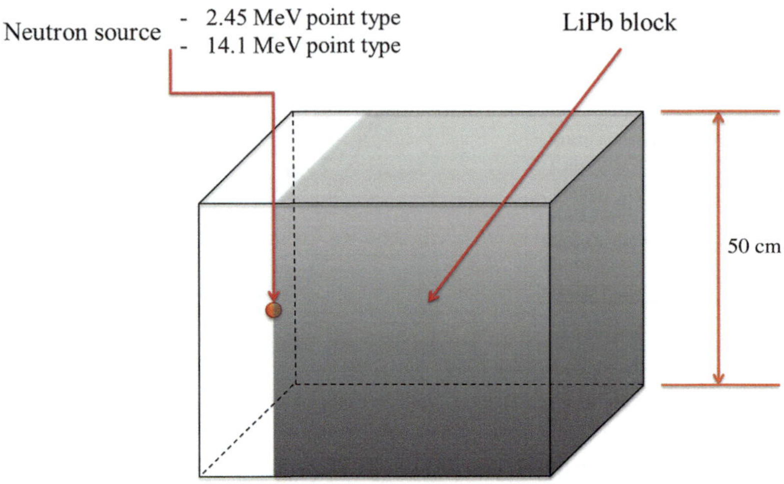

Fig. 3 The schematic image of calculation model 2

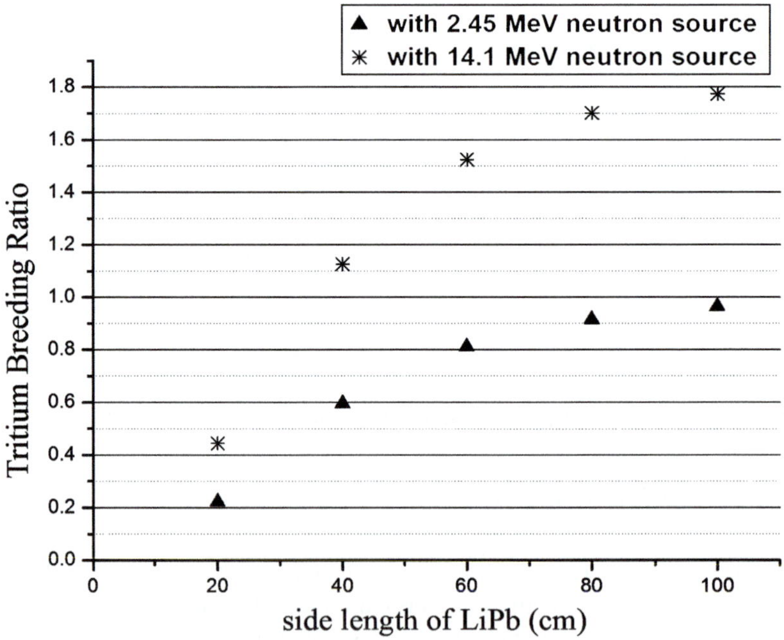

Fig. 4 Tritium production per a neutron with 2.45 MeV, 14.1 MeV

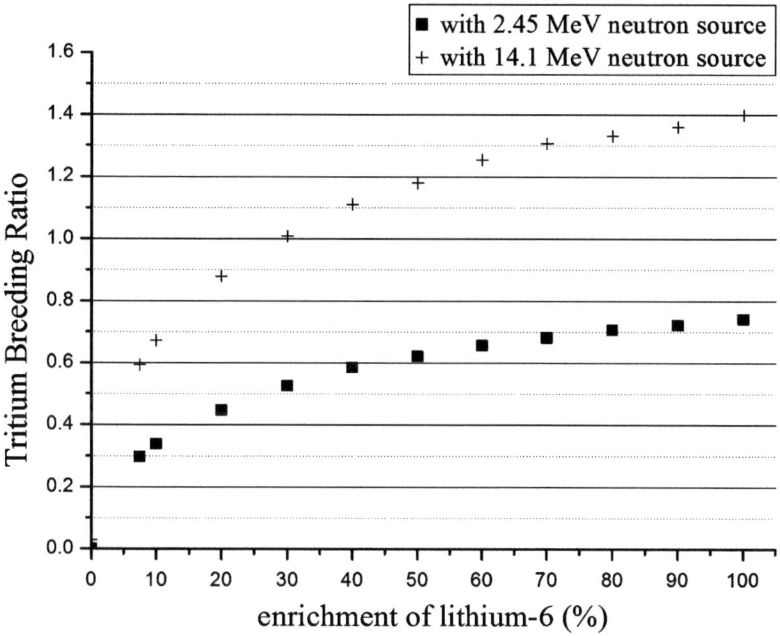

Fig. 5 Tritium production ratio with enrichment of lithium-6(%)

3.2 Calculation Model 2

The distribution of TBR in a module is shown in Fig. 5. As the enrichment of lithium-6 in LiPb increases, the TBR also increases in the calculated module with 50 cm side length. Because the enrichment of lithium-6 in LiPb is large, self-shielding by lithium is expected to occur. The TBR over unity was obtained with the enrichment of lithium-6 more than 30%. According the results, it is expectable to achieve TBR over unity using low enrichment lithium-6 in LiPb.

4 Conclusion

Tritium breeding ratio (TBR) with DT/DD neutron in a simple LiPb blanket module under various calculation conditions were evaluated using Monte Carlo code MCNP. In order to verify the local TBR that it is expected to be greater than 1.43 for considering the coverage of the blanket in the fusion reactor, the suitable size of LiPb block in the integral experiment is between 40 and 60 cm with 14.1 MeV neutron source. The enrichment of lithium-6 greater than 30% is required for the

TBR over unity. Based on the calculation results reported in this study, actual modules for the integral experiment can be designed to verify the total tritium self-sufficiency of the blanket designs.

Acknowledgment The authors were grateful for the support of the Ministry of Education, Culture, Sports, Science and Technology of Japan via "Energy Science in the Age of Global Warming" of Global Center of Excellence (G-COE) program (J-051).

References

1. MCNP-A (1983) General Monte Carlo N-particle transport code, version 5. Report LA-UR-03-1987, Los Alamos
2. Aldama DL, Trkov A (2004) FENDL-2.1: update of an evaluated nuclear data library for fusion applications. IAEA Report INDC(NDS)-467

Analysis of SNIP Algorithm for Background Estimation in Spectra Measured with $LaBr_3$: Ce Detectors

Mohamed Omer, Hani Negm, Ryota Kinjo, Yong-Woon Choi, Kyohei Yoshida, Torgasin Konstantin, Marie Shibata, Kyohei Shimahashi, Hidekazu Imon, Heishun Zen, Toshitada Hori, Toshiteru Kii, Kai Masuda, and Hideaki Ohgaki

Abstract $LaBr_3$:Ce scintillating detectors exhibit excellent properties for γ-ray spectroscopy such as high energy resolution and operation under room temperature as well as MHz counting rates. On the other hand, sever background radiations exist due to the internal contamination of radioactive materials that are very difficult to be avoided during the manufacture. To decrease the effect of these background levels, some analytical techniques, e.g. background subtraction, should be applied. In the present work, we investigate the efficiency of the sensitive nonlinear iterative clipping peak (SNIP) method for background estimation and subtraction. Optimization of the clipping window is discussed for range of energy up to 3 MeV. Enhancement of energy resolution up to 50% was obtained.

Keywords γ-Ray spectra • Internal activity • $LaBr_3$(Ce) • SNIP

1 Introduction

Lanthanum tri-bromide doped with cerium ($LaBr_3$:Ce) is a new comer to the family of scintillating detectors since it has been commercially available from Saint Gobain Crystals under the brand name BrilLanCe380 [1]. The excellent properties of $LaBr_3$:Ce attract many researches and allowed the detector to eventually replace many scintillating detectors. Of these properties are the excellent energy resolution (~3% at 662 keV), high density (5.2 g/cm^3) and very good time resolution

M. Omer (✉) • H. Negm
Institute of Advanced Energy, Kyoto University, Gokasho, Uji, Kyoto 611-0011, Japan

Physics Department, Faculty of Science, Assiut University, Assiut 71516, Egypt
e-mail: omer_sci@iae.kyoto-u.ac.jp; omer_sci@aun.edu.eg

R. Kinjo • Y.-W. Choi • K. Yoshida • T. Konstantin • M. Shibata • K. Shimahashi • H. Imon
H. Zen • T. Hori • T. Kii • K. Masuda • H. Ohgaki
Institute of Advanced Energy, Kyoto University, Gokasho, Uji, Kyoto 611-0011, Japan

T. Yao (ed.), *Zero-Carbon Energy Kyoto 2012*, Green Energy and Technology,
DOI 10.1007/978-4-431-54264-3_27, © Springer Japan 2013

(<300 ps). Therefore, their applications can easily extend to the high energy γ-ray spectroscopy as well as applications that require fast timing. One drawback of $LaBr_3$:Ce is the internal radioactivity due to the existence of the radioactive isotope of lanthanum ^{138}La and also the existence of very small traces of ^{227}Ac which is a chemical homologue of lanthanum. Many researchers investigated the internal activity of $LaBr_3$:Ce crystals in some quantitative manners [2, 3] so that the peaks due to the internal activity can be well recognized in a γ-ray spectrum.

For measurements with high counting rates, the effect of internal radioactivity is negligible, however; for those that the counting rates are very small, further procedures should be applied to enhance spectra measured with $LaBr_3$:Ce in the regions where the internal radioactivity are dominant. These procedures may provide either electronic processing for the pulse generated from the detector [4] or analytical treatment with the pulse recorded in the data acquisition system. In the present work we implement a simple analytical method to enhance the spectra measured with $LaBr_3$:Ce detectors. The method is based on background estimation and subtraction by means of sensitive nonlinear iterative peak clipping algorithm (SNIP) [5, 6]. Fortunately, the abbreviation SNIP means *cut* so we may use the verb *snip* to indicate the SNIP process in the course of this work. In the following, we describe the SNIP algorithm briefly in the Sect. 2. In the Sect. 3, a numerical example is used to evaluate the efficiency of the method and its benefits for both high and low statistical measurements.

2 SNIP Method

The SNIP algorithm aims to estimate the background due to linear and/or nonlinear components under the peaks using the source spectrum. In the algorithm, the minimum of average of the counts at each end of a sampling interval and the channel count at the center of the interval replaces the count in each channel. This sampling interval is arbitrary and is called the clipping window w. The process was repeated for every channel [5] over a number of iterations from one to w. The channel count is minimized to the flat bottom of the peak with increasing the window up to w. Usually the range of counts in a spectrum is very wide (e.g. 1~10^6 counts). Therefore, in order to compress the range of counts and enhance small peaks, the log square root operator (*LLS*) is applied to the source spectrum $T_s(i)$ where i is the channel number, or the bin number for re-binned spectrum. Transformation of *LLS* operator was assigned to another spectrum $T_L(i)$ so that,

$$T_L(i) = \log\left(\log\left(\sqrt{T_s(i) + 1} + 1\right) + 1\right) \quad (1)$$

From $T_L(i)$ we calculate step by step the vectors $T_{L1}(i)$, $T_{L2}(i)$,…, $T_{Lw}(i)$. The new value of channel i is obtained by accepting the minimum of the channel count in and average of the counts at clipping window terminals according to the following equation:

$$T_C(i) = \min\left\{T_L(i), \frac{(T_L(i+w) + T_L(i-w))}{2}\right\} \quad (2)$$

The estimated background B(i) is simply obtained by applying the inverse of the LLS operator.

$$B(i) = LLS^{-1}(T_C(i)) \quad (3)$$

3 Results and Discussion

To evaluate the SNIP algorithm, a measurement was performed using a cylindrical $LaBr_3$:Ce detector with 3.5 in. in diameter and 4 in. in length located at 18 cm from a ^{137}Cs and ^{60}Co radioactive sources. The detector is connected to a proper photomultiplier tube and the pulse arising from radiation interaction with the detector was processed through a digital pulse processing module (Model APU8008 by TechnoAP). The measurement time was around 10 h. Six peaks were used for evaluating of the SNIP algorithm. Three common peaks at 662 keV, 1,173 keV and 1,333 keV arising from ^{137}Cs and ^{60}Co sources are used to test the SNIP algorithm for high statistical measurement. On the other hand, another two peaks at 2,000 keV and 2,614 keV were used to test the algorithm for low statistical measurements. The peak at 1,461 keV is a double peak arising from the superposition of the ^{40} K peak (room background radionuclide) and the sum of 1,436 keV arising from the ^{138}La and the x-ray peaks at 32–37 keV [2]. The peak at 2,000 keV is believed to arise from the α-particles resulting from the decay of ^{227}Ac and its decay chain nuclides. The true energy of the α-particles is from 5 to 7 MeV but because of short range within the crystal, the light yield is corresponding to lower energy. For $LaBr_3$:Ce crystal, the energy of α-particles is around 40% of its true energy [3]. The peak at 2,164 keV is due to the ^{208}Tl isotope that usually exists in nature. Most of the analysis performed in this work was done using the Root Analysis Toolkit [7]. To decrease the number of iterations, the clipping window is the product of the bin width and number of iterations. For example, the clipping window equaling 40 would yield 40 iterations using the channel data and would yield only ten using the re-binned channels to four channels per bin.

Figure 1 shows the spectrum measured with the $LaBr_3$:Ce detector as well as the spectra of the background estimated by SNIP algorithm at different clipping windows (12, 20, 40, 60, 80 and 100 keV). The bin width is 4 keV which corresponds to iterations numbers 3, 5, 10, 15, 20 and 25. As seen from the figure, the lower the clipping window the higher the background estimated. It is a normal behavior since if the clipping window is equal to one which means only *one*

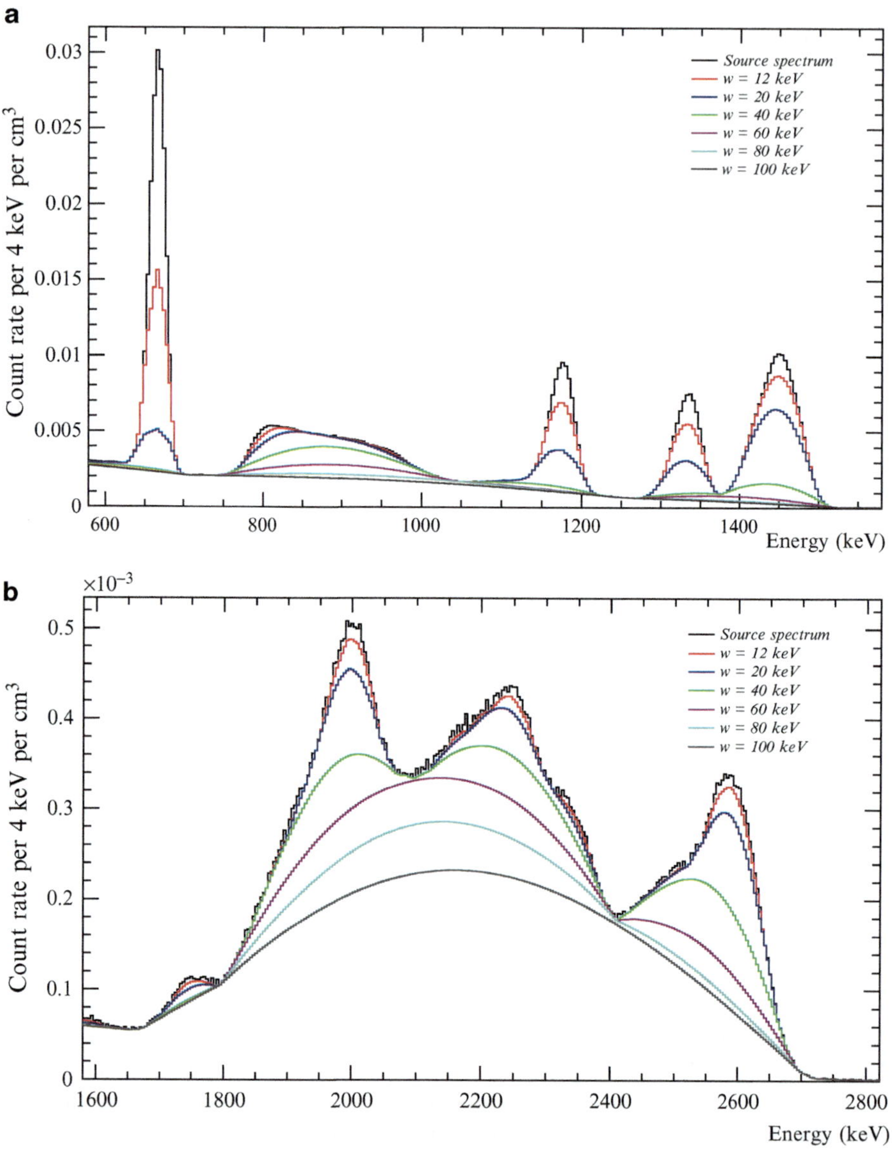

Fig. 1 Source spectrum (in *black*) and the background estimated by SNIP algorithm for different clipping windows. The *upper* graph shows energy range (600–1,600 keV) and the *lower* graph shows the energy range (1,600–2,800 keV)

iteration to minimize the count, the estimated spectrum would be exactly the source spectrum except at the limits i.e. first and last channels.

Figure 2 shows an example of the background subtracted spectrum at clipping window of 40 keV. The linear component resulting from the Compton continuum

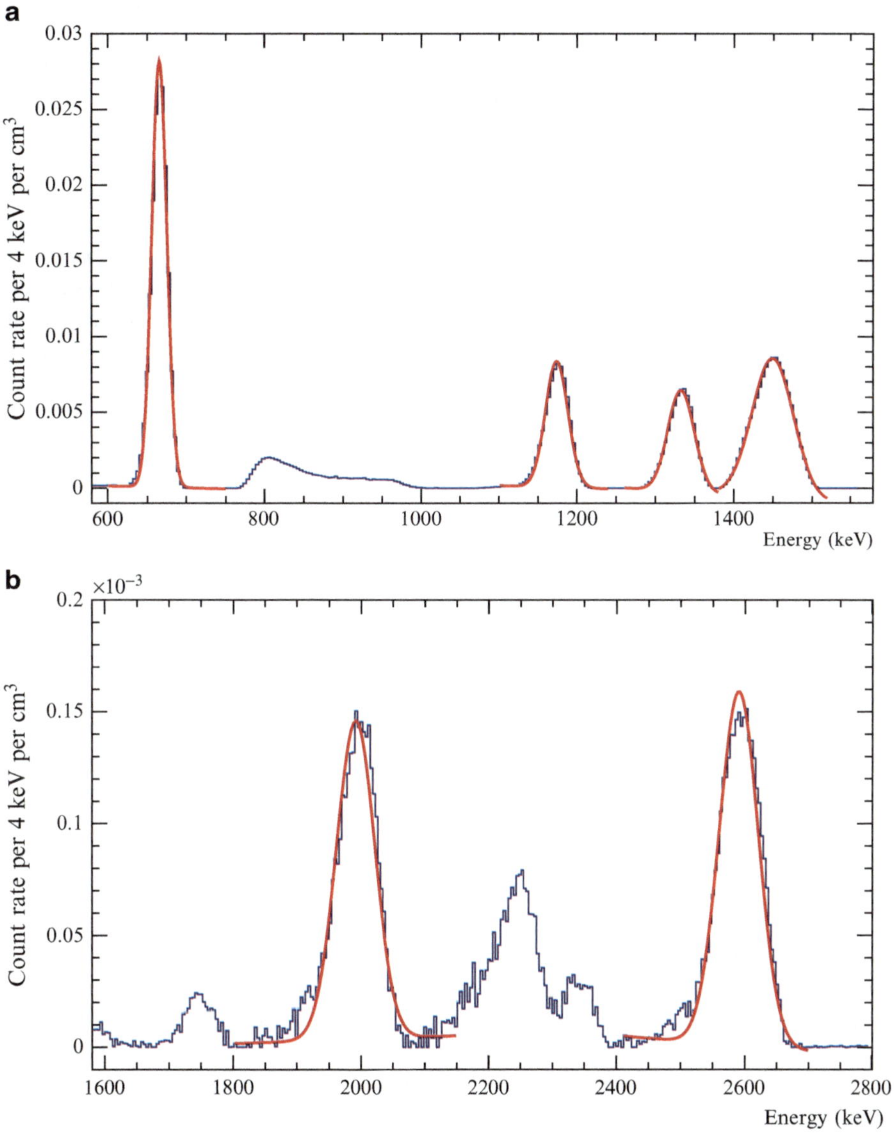

Fig. 2 Background subtracted spectrum at $w = 40$ keV. The *upper* graph shows the low energy range (600–1,600 keV) and the *lower* graph shows the high energy range (1,600–2,800 keV). The *red* curves are Gaussian fits for the peaks

has been completely removed. Furthermore, the complex structure of the spectrum in around 2 MeV has been simplified to great extent and two peaks at 1,750 keV and 2,330 keV are enhanced very much. One more benefit of the SNIP method is that the stability of energy calibration. This can be noted from positions of the peaks before snipping (Fig. 1) and after snipping (Fig. 2).

SNIP algorithm for background estimation and subtraction depends only on one free parameter, the clipping window. In order to assess reliable selection of this parameter, dependence on peaks widths and effect on energy resolution and the net count within the peak were investigated.

We shall define the relative peak count as the net count of a peak after the snipping process divided by the net count of the same peak in the source spectrum at specific clipping window. As shown in Fig. 3 the clipping process causes loss of net peak count. This loss depends on the clipping window applied at specific energy. For example, at $w = 20$ keV the loss in the peak at 662 keV is only 3%, however; the loss at the same clipping window in the peak at 2,614 keV is around 80%. In this example, the clipping window is slightly higher than the full width at half maximum (FWHM) of the peak at 662 keV and it is around one third the FWHM for the peak at 2,614 keV. The same sequence can be estimated for the other peaks. For a simple approximation the net count of the peak is the area under the peak and above the full wave at tenth maximum FWTM while the area under the (FWTM) is a background. Therefore, to minimize the real count loss, the clipping window should evaluate that FWHM $\leq w \leq$ FWTM, however; in some complicated structures of the spectrum w may be slightly lower than the FWHM as shown in the high energy region of the spectrum depicted in Fig. 2. In this case, higher clipping window may not enhance all the peaks shown in the spectrum.

Energy resolution is greatly enhanced by applying the SNIP method on the spectra. Table 1 shows numerical values of the energy resolution in percent for the six peaks at different clipping windows. The energy resolution of the ^{60}Co peaks is below 2% which represents a supreme value for a scintillating detector. The energy resolution of the peak at 1,461 keV is out of any criterion for $LaBr_3$:Ce because it is a double peak. Also, the ^{208}Tl peak is interfered with some peaks of the α-particle contaminants, however; the SNIP method succeeded to isolate this peak.

From the above discussion, the use of the SNIP algorithm for spectra measured with scintillating detectors is important especially for low statistical measurements. Depending on the nature of a measurement, a trade-off should be considered for the required value of resolution versus the amount of loss in the real counts. The loss in real counts is simply a decrease in the detector efficacy so that the trade-off is actually between the energy resolution and detector effeminacy. For very low counting rates experiments, e.g. activation analysis measurements, detection limit is required to claim for detection. The detection limit, in its simplest definition, is a comparison between the net count in a peak and the uncertainty in the background count, the standard deviation of the background count. Therefore, even when the loss in the real count is relatively high, the background of the net spectrum is approaching zero and hence the net count will be much higher than the uncertainty of the background. Therefore, implementation of the SNIP algorithm in a γ-ray spectroscopy software will enhance the detection limit of the whole detection system.

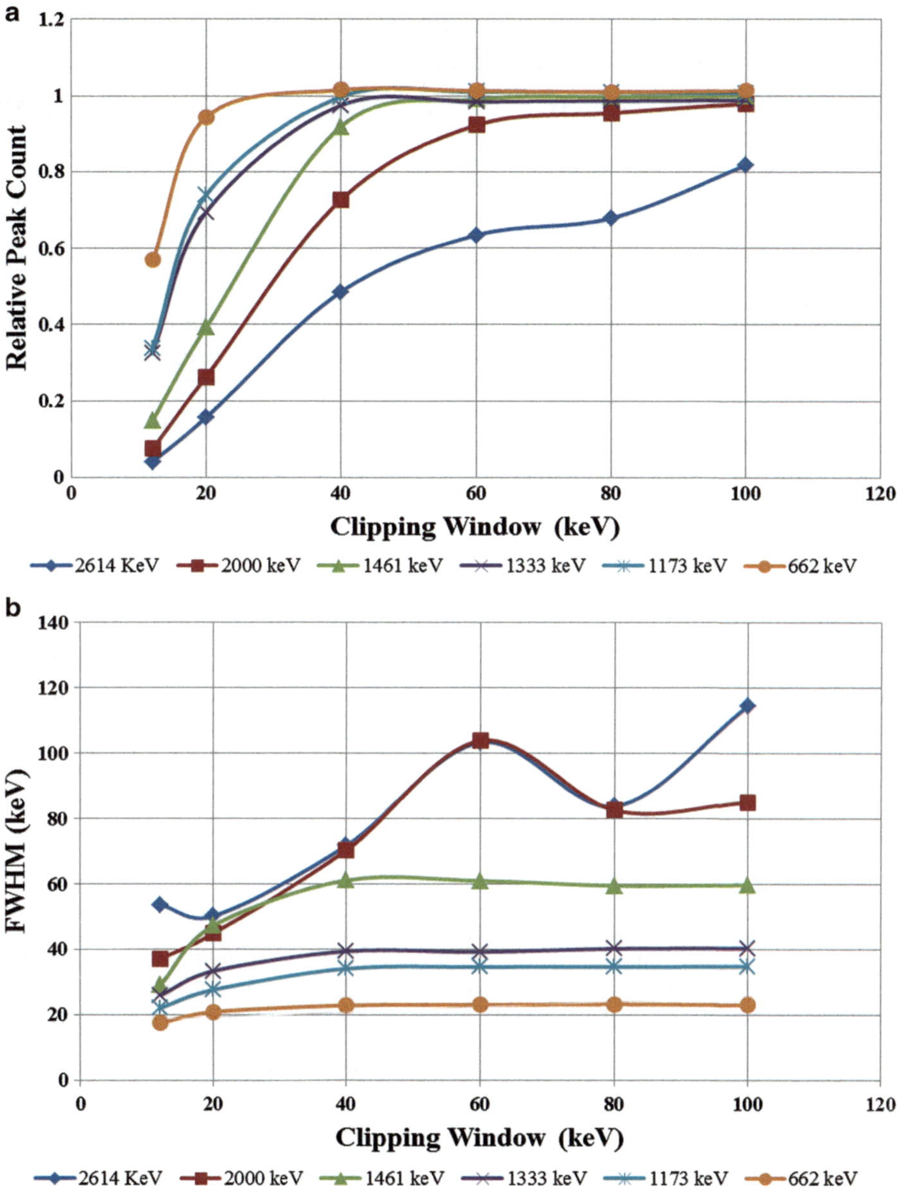

Fig. 3 Relative peak count as a function of the clipping window (*upper*) and FWHM as a function of the clipping window (*lower*)

Table 1 Energy resolution for different energies as affected by the clipping window

	FWHM (%) for different clipping windows					
Energy (keV)	100	80	60	40	20	12
662	3.49	3.52	3.49	3.46	3.16	2.66
1,173	2.96	2.96	2.95	2.90	2.35	1.88
1,333	3.03	3.02	2.94	2.95	2.50	1.95
1,460	4.10	4.08	4.18	4.19	3.24	2.00
2,000	4.26	4.13	5.19	3.52	2.25	1.86
2,614	4.38	3.21	3.95	2.74	1.92	2.05

4 Conclusion

Implementation of SNIP algorithm for the spectra measured with $LaBr_3$:Ce was investigated over a wide range of energies at different clipping windows. A criterion for the clipping window is derived from the analysis of results, FWHM $\leq w \leq$ FWTM. Enhancement up to 50% in the energy resolution is obtained by the aid of the SNIP algorithm. Providing the SNIP algorithm in γ-ray spectroscopy software may be an efficient and fast way to decrease the effect of internal radioactivity and enhance the spectral performance of the detector especially for the low counting rate measurements.

Acknowledgment The authors were grateful for the partial support of G-COE program "Energy Science in the Age of Global Warming" Kyoto University. This work is supported by Special Coordination Funds for Promoting Science and Technology in Japan (Grant No. 066).

References

1. Rosza CM et al (2009) Performance summary BrilLanCeTM scintillators $LaCl_3$ and $LaBr_3$, Saint Gobian Crystals. http://www.detectors.saint-gobain.com/uploadedFiles/SGdetectors/Documents/Technical_Information_Notes/BrilLanCeScintillators-Performance-Summary.pdf
2. Kernan WJ (2006) Self-activity in lanthanum halides. IEEE Trans Nucl Sci 53(1):395–400
3. Nicolini R et al (2007) Investigation of the properties of a 1" × 1" $LaBr_3$:Ce scintillator. Nucl Instrum Methods Phys Res A 582:554–561
4. Crespi FCL et al (2009) Alpha-gamma discrimination by pulse shape in $LaBr_3$:Ce and $LaCl_3$:Ce. Nucl Instrum Methods Phys Res A 602:520–524
5. Ryan CG et al (1988) SNIP, a statistics-sensitive background treatment for the quantitative analysis of pixe spectra in geoscience applications. Nucl Instrum Methods Phys Res B 34:396–402
6. Miroslav M et al (1997) Background elimination method for multidimensional coincidence γ-ray spectra. Nucl Instrum Methods Phys Res A 401:113–132
7. Brun R, Rademakers F (1997) ROOT—an object oriented data analysis framework. Nucl Instrum Methods Phys Res A 389:81–86

A Formulation of Nonlinear Collision Operator for the Monte Carlo Code in Toroidal Plasmas

Yoshitada Masaoka and Sadayoshi Murakami

Abstract A nonlinear collision operator is formulated in order to study the effect of collisions among high-energy particles on their confinements in toroidal plasmas. The Monte Carlo collision operator is derived to be easily implemented to the orbit following code. Using this collision operator, the importance of self collisions of high-energy α particles in a reactor and of high-energy particles by plasma heating, e.g. NBI and ICRF heating can be evaluated.

Keywords α-particle confinement • Monte Carlo method • Nonlinear collision operator • Nuclear fusion • Plasma heating

1 Introduction

In a D-T fusion plasma α particles ($E = 3.5$ MeV) are generated by the fusion reaction and those α particles mainly collide only with electrons because of very high velocity compared to the thermal ion ones. Thus it is considered that the pitch angle scattering is very small for the α particle during the energy slowing down.

On the other hand, the relative velocity between the α particles sometimes becomes very small. In this case they would experience additional pitch angle scattering. Although the density of high-energy particles is much less than that of thermal other ions, the collisions between α-particles would have some effect on pitch angle scattering. It was reported that the nonlinear collisions by each fast α particle has sometimes larger effect than that by other background ions [1].

This nonlinear collision effect may lead to deteriorate the α particle confinement, because of the increase of pitch angle scatterings. Especially, in helical systems,

Y. Masaoka (✉) • S. Murakami
Department of Nuclear Engineering, Kyoto University, Yoshida-honmachi, Sakyo-ku, Kyoto 606-8501, Japan
e-mail: masaoka@p-grp.nucleng.kyoto-u.ac.jp; murakami@nucleng.kyoto-u.ac.jp

T. Yao (ed.), *Zero-Carbon Energy Kyoto 2012*, Green Energy and Technology,
DOI 10.1007/978-4-431-54264-3_28, © Springer Japan 2013

high-energy particle trajectories are complicated due to the three dimensional magnetic configuration and the small increase of pitch angle scattering might affect the high-energy particle confinement significantly.

Therefore, the analysis including the both complicated orbit and the nonlinear collisions are necessary to examine the α-particle confinement in toroidal plasmas. The nonlinear collision operator however has not yet been formulated for the orbit following type of Monte Carlo code. In this paper, we formulate the nonlinear collision operator and the Monte Carlo collision operator which can be easily implemented to the orbit following type codes is derived.

2 The Nonlinear Collision Operator

The nonlinear collision operator was formulated by Killeen [2] and Karney [3], but this operator isn't described in the polar coordinate (v, cosθ) and isn't considered the multi plasma species, where $\cos\theta = v/v_{\parallel}$. The change of the pitch angle by Coulomb collisions is generally described as cosθ and the actual fusion plasma include multi species. Therefore, it is important to formulate the nonlinear collision operator considered the multi plasma species in the polar coordinate (v, cosθ). We consider the Fokker–Plank equation, which is useful for the study of plasma kinetic behaviors. This equation reduces to an equation in time and two velocity dimensions only.

$$\frac{\partial f_a}{\partial t} + v \cdot \nabla f + \frac{q_a}{m_a}(E + v \times B) \cdot \frac{\partial f}{\partial v} - \sum_b C(f_a, f_b) = 0, \tag{1}$$

where f_i is the distribution function of particle species i. v is a drift velocity, q_i and m_i are a charge and mass of species i, E is the electric field and B is the magnetic field. $C(f_a, f_b)$ is the collision term for species a colliding with species b.

Because Coulomb collisions in plasma are typically dominated by small angle scattering, the collision term can be written using the flux as

$$C(f_a, f_b) = -\nabla \cdot \overset{\leftrightarrow}{S}_c^{a/b}, \tag{2}$$

where $\overset{\leftrightarrow}{S}_c^{a/b}$ is the collisional flux in velocity space and is given using Landau collision integral [4] by

$$\overset{\leftrightarrow}{S}_c^{a/b} = \frac{q_a^2 q_b^2}{8\pi\epsilon_0^2 m_a} \ln \Lambda^{a/b} \int \overset{\leftrightarrow}{U}(u) \cdot \left(\frac{f_a(v)}{m_b} \frac{\partial f_b(v')}{\partial v'} - \frac{f_b(v')}{m_a} \frac{\partial f_a(v)}{\partial v} \right) d^3v' \tag{3}$$

where q_i and m_i are the charge and mass of particle species i. ε_0 is the dielectric constant of free space, $\ln \Lambda^{a/b}$ is the Coulomb logarithm, and

$$\overleftrightarrow{U}(u) = \frac{u^2 \overleftrightarrow{I} - uu}{u^3}, \quad u = v - v' . \tag{4}$$

Equation (3) is the useful form of the collision operator for analytical study. But, It is not a convenient form for the numerical implementation. Using an equivalent representation in terms of Rosenbluth potentials, we can rewrite Eq. (3). Using the slightly more convenient notation of Trubnikov, we define two Rosenbluth potentials [5]

$$\phi_b(v') = -\frac{1}{4\pi} \int \frac{f_b(v')}{|v' - v|} d^3v', \tag{5}$$

$$\psi_b(v') = -\frac{1}{8\pi} \int |v - v'| f_b(v') d^3v'. \tag{6}$$

These satisfy Poisson's equations in velocity space

$$\nabla^2 \phi_b(v) = f_b(v), \tag{7}$$

$$\nabla^2 \psi_b(v) = \phi_b(v). \tag{8}$$

$\overleftrightarrow{S}_c^{a/b}$ can be expressed in terms of these potentials as

$$\begin{aligned} \overleftrightarrow{S}_c^{a/b} &= -\overleftrightarrow{D}_c^{a/b} \nabla f_a(v) + F_c^{a/b} f_a(v), \\ \overleftrightarrow{D}_c^{a/b} &= -\frac{4\pi\Gamma^{a/b}}{n_b} \nabla\nabla\psi_b(v), \\ F_c^{a/b} &= -\frac{4\pi\Gamma^{a/b}}{n_b} \frac{m_a}{m_b} \nabla\phi_b(v), \end{aligned} \tag{9}$$

where $\overleftrightarrow{D}_c$ and F_c are *a diffusion tensor* and *a friction force*, respectively, and

$$\Gamma^{a/b} = \frac{n_b q_a^2 q_b^2 \ln \Lambda^{a/b}}{4\pi\epsilon_0^2 m_a^2}. \tag{10}$$

Because of a singular point at $v' = v$ it is difficult to calculate ϕ_b numerically. However, there is a useful method for the calculation of the Rosenbluth potentials. The method needs the expansion of the distribution function f by Legendre harmonics $P_l(v, \cos\theta)$, where $\cos\theta = v_{\|}/v$. The Legendre expansion of the distribution function $f(v, \cos\theta)$ is written as

$$f(v,\cos\theta) = \sum_{l=0}^{\infty} f^{(l)}(v) P_l(\cos\theta), \tag{11}$$

where

$$f^{(l)}(v) = \frac{2l+1}{2} \int_0^{\pi} f(v,\cos\theta) P_l(\cos\theta) \sin\theta d\theta. \tag{12}$$

The Legendre polynomials are evaluated using the recurrence relation [6]

$$\begin{aligned} P_0(\mu) &= 1, \\ (l+1)P_{l+1}(\mu) &= (2l+1)\mu P_l(\mu) - lP_{l-1}(\mu) \end{aligned}.$$

Using this decomposing distribution function f, the Rosenbluth potentials for each species b are written as

$$\phi_b^{(l)}(v) = -\frac{1}{2l+1}\left[\int_0^v \frac{v'^{l+2}}{v^{l+1}} f_b^{(l)}(v')dv' + \int_v^{\infty} \frac{v^l}{v'^{l-1}} f_b^{(l)}(v')dv'\right], \tag{13}$$

$$\begin{aligned} \psi_b^{(l)}(v) = \frac{1}{2(4l^2-1)} &\left[\int_0^v \frac{v'^{l+2}}{v^{l-1}}\left(1 - \frac{(l-\frac{1}{2})v'^2}{(l+\frac{3}{2})v^2}\right) f_b^{(l)}(v')dv' \right. \\ &\left. + \int_v^{\infty} \frac{v^l}{v'^{l-3}}\left(1 - \frac{(l-\frac{1}{2})v^2}{(l+\frac{3}{2})v'^2}\right) f_b^{(l)}(v')dv'\right]. \end{aligned} \tag{14}$$

Thus the total Rosenbluth potentials are written as

$$\phi_a(v,\lambda) = \sum_{l=0}^{\infty} \sum_b \frac{m_a + m_b}{m_b} \phi_b^{(l)}(v) P_l(\lambda), \tag{15}$$

$$\psi_a(v,\lambda) = \sum_{l=0}^{\infty} \sum_b \psi_b^{(l)}(v) P_l(\lambda). \tag{16}$$

Since the change of the pitch angle by collisions is generally described as and proportionate to $\lambda = \cos\theta$, we describe the drift velocity vector $\mathbf{v} = (v, \cos\theta)$. In the polar coordinate (v, λ), $\nabla\nabla\ \psi$ and $\nabla\ \phi$ involving the friction force vector F and diffusion tensor $\overleftrightarrow{D}$ in Eq. (9) become

$$(\nabla\nabla\psi)_{vv} = \frac{\partial^2\psi}{\partial v^2}, \tag{17}$$

$$(\nabla\nabla\psi)_{\lambda\lambda} = \frac{1-\lambda^2}{v^2}\left(\frac{1-\lambda^2}{v^2}\frac{\partial^2\psi}{\partial\lambda^2} - \frac{\lambda}{v^2}\frac{\partial\psi}{\partial\lambda} + \frac{1}{v}\frac{\partial\psi}{\partial v}\right), \tag{18}$$

$$(\nabla\nabla\psi)_{v\lambda} = (\nabla\nabla\psi)_{\lambda v} = -\frac{1-\lambda^2}{v^2}\left(\frac{\partial^2\psi}{\partial v\partial\lambda} - \frac{1}{v}\frac{\partial\psi}{\partial\lambda}\right), \tag{19}$$

$$(\nabla\phi)_v = \frac{\partial\phi}{\partial v}, \tag{20}$$

$$(\nabla\phi)_\lambda = \frac{\partial\phi}{\partial\lambda}. \tag{21}$$

And the friction force vector F and diffusion tensor $\overleftrightarrow{D}$ become

$$\overleftrightarrow{D} = -\frac{4\pi\Gamma^{a/b}}{n_b}\begin{pmatrix}(\nabla\nabla\psi)_{vv}(\nabla\nabla\psi)_{v\lambda}\\(\nabla\nabla\psi)_{\lambda v}(\nabla\nabla\psi)_{\lambda\lambda}\end{pmatrix}, \tag{22}$$

$$F = -\frac{4\pi\Gamma^{a/b}}{n_b}\frac{m_a}{m_b}\begin{pmatrix}(\nabla\phi)_v\\(\nabla\phi)_\lambda\end{pmatrix}. \tag{23}$$

3 Monte Carlo Collision Operator

In order to analyze the confinement of high-energy ions in the presence of collisions, it is very useful to use the code based on the Monte Carlo method. Therefore, it is necessary to derive the Monte Carlo collision operator from Fokker–Planck type equation.

Monte Carlo operators which represent a random walk in $I(v, \lambda)$ with a time step Δt are now given by changing the values of the invariants I^i at the nth step to those at the next step, as follows

$$I^i_{n+1} = I^i_n + \Delta I^i, \tag{24}$$

where the quantities ΔI^i are stochastic values. The expectation values (the friction force) and covariances (the diffusion) of them are given by

$$E(\Delta I^i) = \left[F^i + \frac{1}{v^2}\frac{\partial}{\partial I_j}\left(v^2\overleftrightarrow{D}^{ij}\right)\right]\Delta t, \tag{25}$$

$$C(\Delta I^i, \Delta I^j) = 2\overset{\leftrightarrow ij}{D} \Delta t. \tag{26}$$

In general, the covariance matrix $C(\Delta I^i, \Delta I^j)$ is symmetric and positive definite. Thus, it can be diagonalized, i.e., a unitary matrix M such as

$$C(\Delta I^i, \Delta I^j) = M^{ik} \Lambda^{kl} M^{jl} \tag{27}$$

exists, where $\Lambda^{kl} = \mathrm{diag}(\Lambda_1, \Lambda_2)$ is the diagonal matrix of the . Because the these eigenvalues of matrix C are positive, we can use matrix $L^{kl} = diag(\sqrt{\Lambda_1}, \sqrt{\Lambda_2})$, write the covariance matrix as

$$C(\Delta I^i, \Delta I^j) = M^{ik} L^{km} L^{lm} M^{jl}. \tag{28}$$

From this relation, as well as Eqs. (25) and (26), ΔI^i is expressed as

$$\Delta I^i = E(\Delta I^i) + M^{ik} L^{kl} \xi^l, \tag{29}$$

where ξ^l is $\xi = \pm 1$ and this sign is chosen at random. This completes the construction of the Monte Carlo operator.

4 Conclusion

The nonlinear collision effect between high-energy α particles may lead to deterioration of the α particle confinement, owing to the increase of the pitch angle scattering. Thus the analysis including both complicated orbits and nonlinear collision effects are necessary to examine the α-particle confinement. However, this nonlinear collision operator has not been formulated for the Monte Carlo type code.

We have formulated the nonlinear collision operator in order to estimate the collision effect between high-energy α particles on their confinements. We have derived the Monte Carlo collision operator to be easily implemented to the orbit following type code. Using this collision operator, we can evaluate the importance of self collisions of high-energy α particles in a reactor and of high-energy particles by plasma heating, e.g. NBI and ICRF heating.

Acknowledgements The authors were grateful for the support of the Ministry of Education, Culture, Sports, Science and Technology of Japan via "Energy Science in the Age of Global Warming" of Global Center of Excellence (G-COE) program (J-051).

References

1. Okano K, Umeda N, Kamada Y et al (2001) Non-linear increment of beam stopping cross-section by circulating fast ions. In: Proceedings of the 28th EPS Conference Control. Fusion Plasma Physics, pp 809–812
2. Killeen J, Kerbel GD, McCoy MG, Mirin AA (1986) Computational methods for kinetic models of magnetically confined plasmas. Springer, New York
3. Karney CFF (1986) Comput Phys 4:183
4. Landau LD (1936) Phys Z Sowjet 10:154
5. Rosenbluth N et al (1957) Fokker-Planck equation for an inverse-square force. Phys Rev 107:1
6. Abramowitz M, Stegun IA (1965) Handbook of mathematical functions. Dover, New York

References

The Rheological Characters of Surfactant Viscoelastic Solutions at Low Shear Rate

Li-Fang Jiao, Tomoaki Kunugi, and Feng-Chen Li

Abstract Rheological properties of dilute surfactant solutions at low shear rate are investigated using a cone viscometer. The plateau viscosity changing with the shear rate can be separated to three regimes: in regime 1, the viscosity decreases with increase of the shear rate while the shear rate is below a threshold shear rate; in regime 2, the shear rate is above the threshold shear rate and the viscosity tends to a little bit increase as the shear rate is increased; In regime 3, the viscosity gradually decreases again with increase of the shear rate. The influencing factors on its viscosity contain concentration, temperature and the weight ratio of counter ions. Through rotating the solutions at a proper shear rate, a significant change of their viscosity at low shear rate was found: it is characterized by large amplified periodic vibration of the tested viscosity. This large vibration is thought to result from the macro worm-like micelles in the viscoelastic surfactant solutions.

Keywords Micelle • Surfactant • Viscoelastic • Viscosity

1 Introduction

Surfactants are industrially ubiquitous in both chemical and consumer domains and have provided a rapidly growing area of scientific research. They are used as detergents, cleaning agents, emulsifiers in food, pharmaceuticals, and cosmetics. Surfactant molecules in solution can self-assemble into aggregates (so-called micelles) which their immiscible parts clump together and let solution express

L.-F. Jiao • T. Kunugi (✉)
Graduate School of Engineering, Kyoto University, Yoshida Honmachi, Sakyo, Kyoto, Japan
e-mail: kunugi@nucleng.kyoto-u.ac.jp

F.-C. Li
School of Energy Science and Engineering, Harbin Institute of Technology, Harbin, China

T. Yao (ed.), *Zero-Carbon Energy Kyoto 2012*, Green Energy and Technology, 261
DOI 10.1007/978-4-431-54264-3_29,

non-Newtonion property which can highly influence flow characteristics in practical applications [1]. The shape and size of micelle depend on the molecular structure of the surfactant, nature of solvent and additives, and their molar composition [2–4]. Specifically, the ionic surfactant molecules form spherical micelles when their concentration exceeds the critical micelle concentration (CMC) [5]. The micellar aggregates can grow anisotropically under appropriate conditions, such as proper shear action, changing their shapes from spheres to rods or highly flexible wormlike aggregates [6, 7]. In addition, the structural transitions are highly associated with the presentation of count-ionic salts, which can reduce the repulsions between micelle-head groups [3, 7].

In this work, we examined experimentally aqueous solutions of a cationic surfactant (Cetytrimethylammonium Chloride, CTAC henceforth) in the presence of strongly binding salicylic counter ions (NaSal hereafter), which expressed high viscoelasticity at proper conditions and is proved to be good drag reducing agents [8]. Several studies were performed on the relationship of the micro-structural change and rheological properties of wormlike micellar solutions [9, 10]. In the dilute regime, the solutions behave as a real sol with a rather low viscosity. As the shear flow applied, the wormlike micellar solution exhibits anomalous behavior, such as a dramatic increase in viscosity above a critical shear rate. Above a well-defined threshold value of the velocity gradient, a supramolecular structure can be formed from the micellar aggregates, which behaves like a spring. It has been suggested that these phenomena are associated with aggregation of micelles induced by the shear flow [9–14]. In the present work, we considered the rheological responses and flow induced microstructure of dilute wormlike micellar solutions by measuring the solution viscosity under shear flow. The developing process of flow induced microstructures was investigated through analyzing the changes of apparent viscosity with time.

2 Experimental

CTAC and NaSal were purchased from Nacalai Tesque, Co. Ltd. in Japan. Highly de-ionized water was used as a solvent. The CTAC concentration was set in the range from 100 to 1,000 ppm and the NaSal concentration was varied with the weight ratio, $R = W_{CTAC}/W_{NaSal}$, was changed from 0.5 to 2.0. Before measurements were taken, samples were kept for 5 days at least to attain equilibrium.

Rheological properties of CTAC solutions were measured using BROOKFIELD LVDV-II + Pro (*CPE-40 cone*) viscometer and RE-85 (*stand cone 1.34*R24*) viscometer. Later is mainly used to test the temperature influence. Applied shear rate was varied from 5 to 999 s^{-1}.

3 Results and Discussion

3.1 Salicylic Counter-Ions and Concentration Dependence

It was proposed to get a stronger micelle structure by adding equivalent weight concentration of NaSal to CTAC solutions [6]. Here, three weight ratios, R = 1:0, 1:0.5, 1:1 and 1:2, were chosen to investigate the salicylic counter-ion effect. While R = 1:0, the pure 500 ppm CTAC aqueous solution expresses the Newtonian fluid behavior and its viscosity is similar to the pure water's. It is proposed that, in this case, the micellar shape should be sphere, and the repulsion between micelle head groups is too strong to further aggregate as a supramolecular. The addition of NaSal salt accelerates the aggregation of micelles and let the solution exhibit a non-linear rheological behavior with shear rate, which is expressed by the significant increase of viscosity here as shown in Fig. 1. With increase of the acting shear rate, viscosity sharply decreases at low shear rate (regime 1), and then tends to a little bit increase while the shear rate arrives a threshold value (regime 2), and finally further decreases (here the maximum shear rate is 383 s^{-1} and the corresponding solution viscosity is larger than 2.9 mPa s, which is nearly three times of water's viscosity, therefore 383 s^{-1} is not enough to destroy the shear induced wormlike structures) (regime 3). Comparing these three weight-ratio cases, the threshold shear-rate value is largest when the weight ratio equals to unity, R = 1:2 case takes the second place, and its value is smallest in R = 0.5 case. The larger threshold shear rate means more difficult to form wormlike structures, in other word, the micellar structures are stronger. This result further proved that the optimum weight ratio of NaSal and CTAC was 1:1. As a result, we chose R = 1:1 CTAC/NaSal solutions as a working fluid.

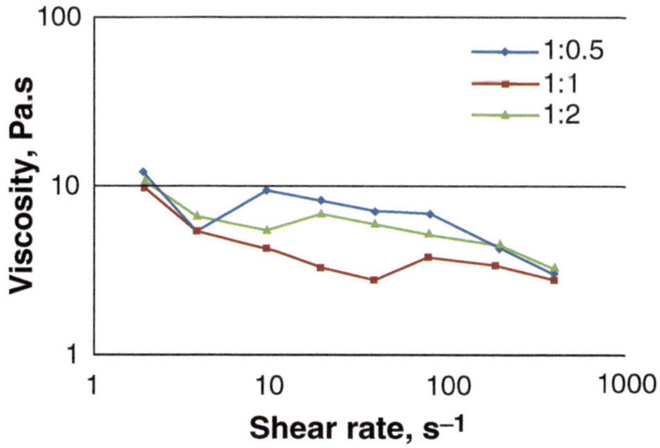

Fig. 1 Effect of weight ratio of CTAC to NaSal on the shear viscosity for 500 ppm CTAC solutions, T = 25°C

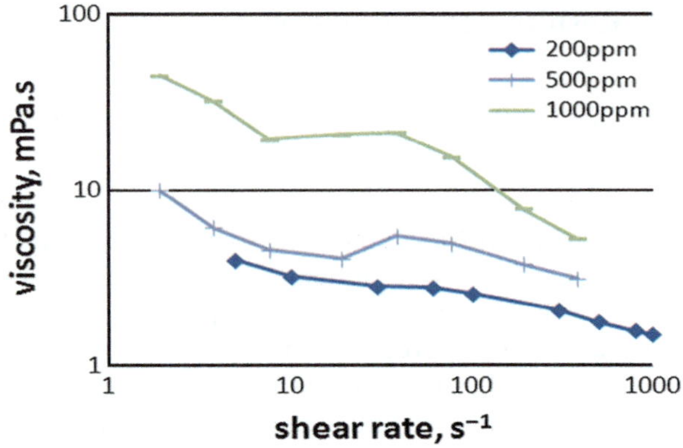

Fig. 2 Effect of concentration on the shear viscosity for R = 1 CTAC/NaSal solutions, T = 25°C

In order to investigate the concentration effect, 200 ppm, 500 ppm and 1,000 ppm CTAC/NaSal were tested while the weight ratio R was fixed at 1:1 and temperature was controlled at T = 25 ± 0.5°C. The results are reproduced in Fig. 2. An apparent increase of the viscosity can be found with increase of the CTAC/NaSal concentration through the further aggregation of micelles.

4 Temperature Dependence

From the standpoint of energy conservation, higher temperature can reduce the strength of micelles by enhancing the inner activity of micelles. As shown in Fig. 3, the changes of viscosity of 500 ppm CTAC/NaSal expressed nearly Newtonian fluid behavior at T = 40°C, while strong non-Newtonian fluid property is shown at T = 15°C. Based on the results in Fig. 3, it is possible to proposed that CTAC/NaSal solution is very sensitive to temperature variation and with the increase of solution temperature, the viscoelasticity becomes weaker and weaker until totally disappear.

5 Shear Rate and Time Dependence

In the above tests, the maximum acting shear rate is 383 s^{-1}. Although this shear rate is not so large to destroy the shear induced wormlike structure as mentioned in 3.1, it is still not clear how large it can elongate the shear induced structure up to now. The wormlike structure likes a spring, and its stretching length increases with the increase of the acting shear stress until the wormlike structure to be broken: the maximum acting shear rate is further increased to 999 s^{-1}. CTAC/NaSal was measured once the shear rate increased from 5 to 999 s^{-1} and was then back to 5 s^{-1}

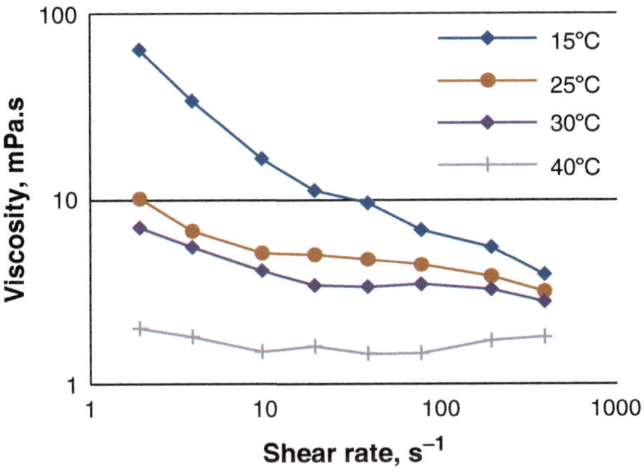

Fig. 3 Temperature effect on the shear viscosity for 500 ppm, R = 1 CTAC/NaSal solutions

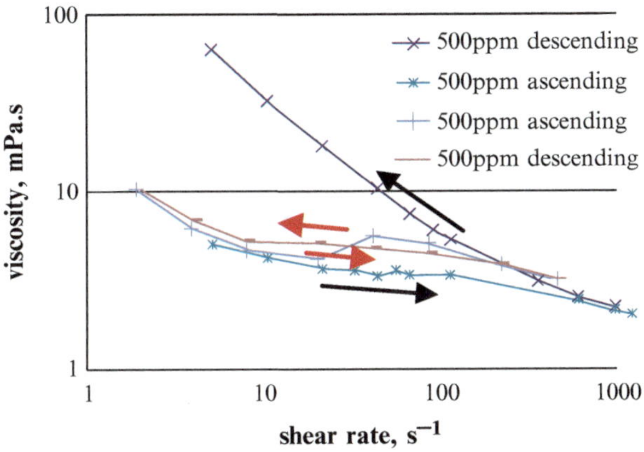

Fig. 4 Shear rate dependence of the shear viscosity for 500 ppm, R = 1 CTAC/NaSal solutions, T = 25°C

at concentration = 500 ppm, R = 1:1 and T = 25°C. One of the surprising findings is the plateau viscosity was significantly enlarged while the shear rate was back to 5 s^{-1} from 999 s^{-1} comparing to the shear rate gradually ascending case, as shown in Fig. 4. As to find out the detail reason of this phenomenon, further observation was done on the developing process of the viscosity in the shear flow of "500 ppm, R = 1, T = 25°C CTAC/NaSal". The results are shown in Fig. 5.

Below the shear rate is $40s^{-1}$, the plateau viscosity exhibits the periodical vibration regularly after a short reaction time Δt_1 as shown in Fig. 5a, and the plateau viscosity decrease with increase of the shear rate. It is speculated that the micellar aggregates change their shapes from sphere to ellipsoid through reforming at low shear rate cases. At $50s^{-1}$, the viscosity further increases after the reaction time, Δt_1, and then exhibits the periodical vibration regularly after a stretching time, Δt_2, with a larger period (this means the micelles are stretched longer) as shown in Fig. 5b. Therefore, it is deduced that the micelles are successfully induced from elliptical

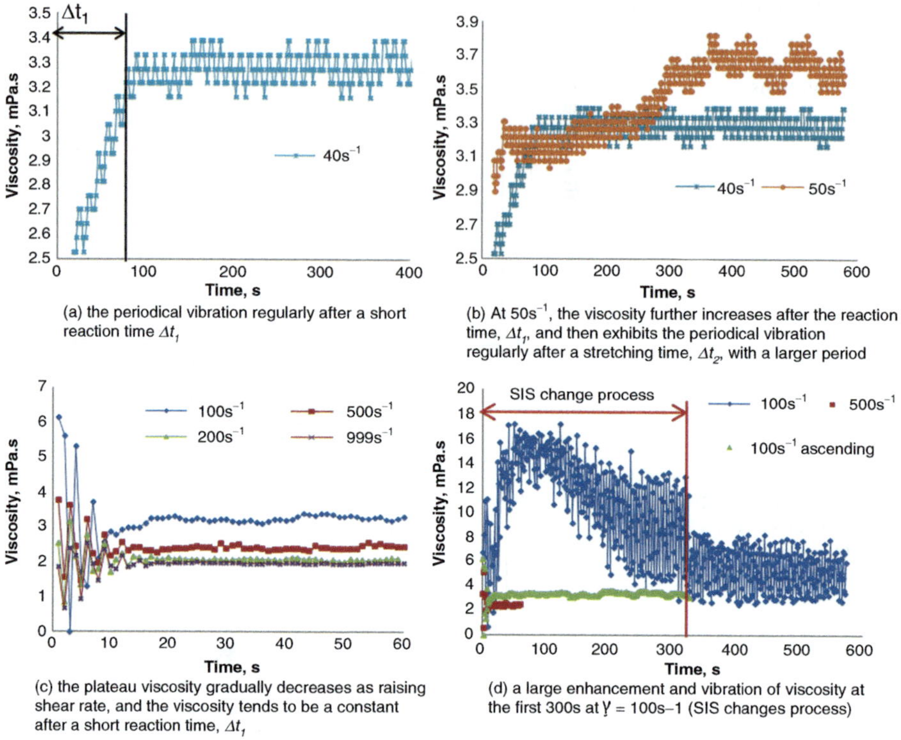

(a) the periodical vibration regularly after a short reaction time Δt_1

(b) At 50s^{-1}, the viscosity further increases after the reaction time, Δt_1, and then exhibits the periodical vibration regularly after a stretching time, Δt_2, with a larger period

(c) the plateau viscosity gradually decreases as raising shear rate, and the viscosity tends to be a constant after a short reaction time, Δt_1

(d) a large enhancement and vibration of viscosity at the first 300s at γ = 100s–1 (SIS changes process)

Fig. 5 Shear rate and time dependence of 500 ppm, R = 1 CTAC/NaSal solutions, T = 25°C

shape to rod and wormlike shape. After that, the plateau viscosity gradually decreases as raising shear rate, and the viscosity tends to be a constant after a short reaction time, Δt_1, as shown in Fig. 5c. While the shear rate is gradually back to 5 s^{-1} from 999 s^{-1}, a large enhancement and vibration of viscosity are found at the first 300 s at $\gamma = 100$ s^{-1} and then the vibration becomes regular, stable and periodical, but the value of plateau viscosity is still larger than its value in the shear rate ascending case. It can be expected that the elongated micelles are aligned to flow direction while the shear stress equals to the elastic normal force in micelles, such as the shear rate larger than 100 s^{-1} cases. While the elastic force in a micelle is larger than the shear stress acting on it, at the beginning of the shear stress action, the coiled wormlike micelles are stretched along the flow direction. The elastic force gradually increases with increase of micelle length until its value is large enough to resist the shear stress action, and recovers the elongated micelle to coiled wormlike structure, then begins the next elongation cycle and be curling up behaviors, which corresponds to the large vibration of tested viscosity, as shown in Fig. 5c, d and Fig. 6a. The vibration amplitude of the viscosity indirectly responds the strength of elasticity of the solution. Further analysis suggested that the vibration amplitude of the viscosity was an exponential function of the acting shear rate as shown in Fig. 6b.

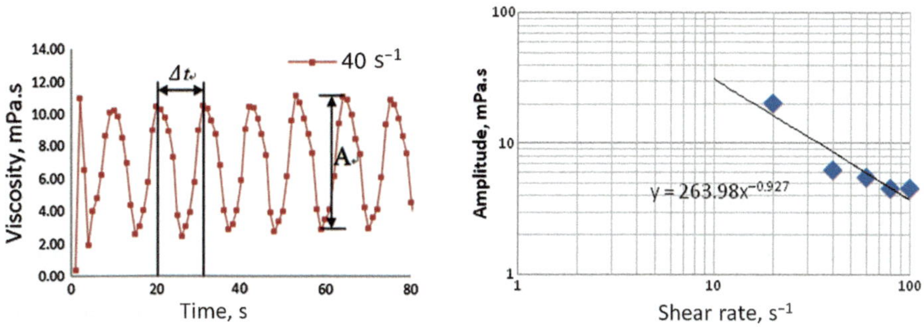

(a) Periodical vibration of the shear viscosity (b) the vibration amplitude of the viscosity vs. acting shear rate

Fig. 6 Periodical vibration of the shear viscosity of 500 ppm, CTAC/NaSal solutions, T = 25°C

6 Summary

The rheological behavior of CTAC/NaSal solution depends strongly on the concentration and solution temperature of CTAC. The plateau viscosity changing with the shear rate can be separated to three regimes: in regime 1, the viscosity decreases with increase of the shear rate while the shear rate is smaller than a critical shear rate; in regime 2, which occurs for the shear rate larger than a critical shear rate, the viscosity tends to a bit increase as the shear rate is raised; In regime 3, the viscosity decreases again with increase of the shear rate. Rotating the solutions at a proper shear rate, a significant change of their viscosity at low shear rate was found, which is characterized by a large amplified periodic vibration of the test viscosity. It can be said that this large vibration is the result from the interaction of elastic force in micelles and the acting shear stress.

Acknowledgment This work was partly supported by an "Energy Science in the Age of Global Warming" of Global Center of Excellence (G-COE) program (J-051) of the Ministry of Education, Culture, Sports, Science and Technology of Japan.

References

1. Zakin JL, Lu B, Bewersdorff HV (1998) Surfactant drag reduction, an invited review article. Rev Chem Eng 14:255–320
2. Goyal PS, Dasannacharya BA et al (1991) Shapes and sizes of micelles in CTAB solutions. Phys B 174:196–199
3. Ikeda S, Ozeki S, Hayashi S (1980) Size and shape of charged micelles of ionic surfactants in aqueous salt solutions. Biophys Chem 11(3–4):417–423
4. Chevalier Y, MeliSa F, Dalbiezs JP (1992) Structure of zwitterionic surfactant micelles: micellar size and intermicellar interactions. J Phys Chem 96:8614–8619

5. Ruckenstein E, Nagarajan R (1975) Critical micelle concentration. A transition point for micellar size distribution. J Phys Chem 79(24):2622–2626
6. Alargova RG, Danov KD, Petkov JT et al (1997) Sphere-to-rod transition in the shape of anionic surfactant micelles determined by surface tension measurements. Langmuir 13:5544–5551
7. Lin Z, Cai JJ, Scriven LE et al (1994) Spherical-to-wormlike micelle transition in CTAB solutions. J Phys Chem 98(23):5984–5993
8. Hadri F, Guillou S (2010) Drag reduction by surfactant in closed turbulent flow. Int J Eng Sci Tech 2(12):6876–6879
9. Kim WJ, Yang SM (2000) Flow-induced silica structure during in situ gelation of wormy micellar solutions. Langmuir 16:4761–4765
10. Kim WJ, Yang SM (2000) Microstructure and rheological response of aqueous CTAB solutions in the presence of benzyl additives. Langmuir 16:6084–6093
11. Liu CH, Pine DJ (1996) Shear-induced gelation and fracture in micellar solutions. Phys Rev Lett 77:2121–2124
12. Hu YT, Boltenhagen P, Pine DJ (1998) Shear thickening in low-concentration solutions of wormlike micelles. I. Direct visualization of transient behavior and phase transitions. J Rheol 42:1185–1208
13. Humbert C, Decruppe JP (1998) Stress optical coefficient of viscoelastic solutions of CTAB and KBr. Colloid Polym Sci 276:160–168
14. Choi DG, Kim WJ, Yang SM (2000) Shear-induced microstructure and rheology of cetylpyridinium chloride/sodium salicylate micellar solutions. Korea Aust Rheol J 12:143–149

Characteristics of Velocity Fields and Polymers' Elongation in Elastic Turbulent Flow

Hong Na Zhang, Tomoaki Kunugi, and Feng Chen Li

Abstract The present paper describes the elastic turbulent flow numerically generated in a parallel plate channel. On the basis of the numerical results, the characteristics of the velocity fields, the polymer elongations and the underlying interaction process were investigated to get insight into this phenomenon. It is indicated that increasing the Weissenberg number, the flow oscillations become stronger and further intermittent with larger-time-scale flow structures. The transversal velocity was generated in the form of quadrupolar patterns localized in the regions with strong polymers' feedback. Besides, the elastic instability vortex-tube-like structures (regions dominated by rotating motion) are formed mainly surrounding the regions where the polymers are strongly deformed.

Keywords Elastic turbulent flow • Polymers and flow interaction • Polymers elongation

1 Introduction

Due to the strong flexibility of the polymer molecule's long chains, the addition of small amount of polymers into the Newtonian solvent could significantly modify the flow properties and lead to two nonlinear effects in the viscoelastic fluid flow: the nonlinear inertial and nonlinear elastic effects which are not necessary to coexist [1]. The characteristics of the viscoelastic fluid flow are strongly dependent on the strength of the feedback generated by the polymers. Therein, the strength is related to polymers' behaviors or elongation rate, determined by the Weissenberg

H.N. Zhang • T. Kunugi (✉)
Graduate School of Engineering, Kyoto University, Kyoto, Japan
e-mail: kunugi@nucleng.kyoto-u.ac.jp

F.C. Li
Harbin Institute of Technology, School of Energy Science and Engineering, Harbin, China

T. Yao (ed.), *Zero-Carbon Energy Kyoto 2012*, Green Energy and Technology,
DOI 10.1007/978-4-431-54264-3_30, © Springer Japan 2013

number Wi, which is defined as the product of polymers relaxation time λ and the characteristic shear rate of the flow. In other words, the Wi indicates the degree of anisotropy or orientation generated by the deformation of the polymer in the simple steady shear. If Wi $\ll 1$, the polymers are relaxed by their elasticity much faster as compared with the flow stretching, and therefore the polymers are always in the coiled state and passively embodied in the flow. In this case, the feedback effect on the flow can be neglected. However, if further increasing the Wi to the order of unity or larger, the strong feedback will be generated due to the faster stretching from the flow and growth of its length. For example, the substantial reduction of turbulent drag force is due to this feedback effect [2]. Another striking and intriguing effect on the flow is the occurrence of flow instability and even turbulent-like regime at an arbitrary low Reynolds number (Re) in the upper limit of Wi, named as the elastic instability and turbulence ([3–6], and references therein). Under the same Re the Newtonian fluid flow is definitely in the laminar regime. The possibility of the "elastic turbulence" occurrence has a great potential in improving mixing efficiency in the micro-scale areas [6–9].

Comprehensive researches on the elastic turbulence have been carried out through the experiments done by Steinberg group [10–13] and limited to the geometry with a local curvature such as the swirling flows between two disks, Taylor-Couette flows between two coaxial cylinders, and also micro-scale flows in the curvilinear channels. Therein, the "hoop stress" (defined as the ratio of the local first normal stress difference and the curvature) is regarded as the dominated mechanism. Because of the degradation of the polymers the elastic instability and turbulence cannot be experimentally observed in the flow without a local curvature. Besides, the difficulty of measuring polymers' behaviors in the elastic turbulence also limits our current understanding of its mechanism. In this sense, numerical simulation including the consideration of polymers' elongation through a proper viscoelastic fluid model is a useful tool to solve this problem. Actually, several numerical simulations have been carried out, such as Taylor-Couette flow [14], Kolmogorov flow [15–18] and Stokesian flow [19], the flow in three-dimensional (3D) curvilinear channel [20] of viscoelastic fluids, in order to elaborate the flow pattern, instability or turbulence inception mechanism and mixing enhancement therein. Moreover, some simulations [15–18] also confirm that it is possible to generate the elastic instability and turbulence in the geometry without local curvature. The current work further enriches the database of the elastic turbulence in the geometry without local curvature but with wall effects as compared with Kolmogorov flow [15–18].

2 Numerical Methods

3D direct numerical simulations (DNSs) of the viscoelastic fluid flow were carried out in a parallel plate channel illustrated in Fig. 1.

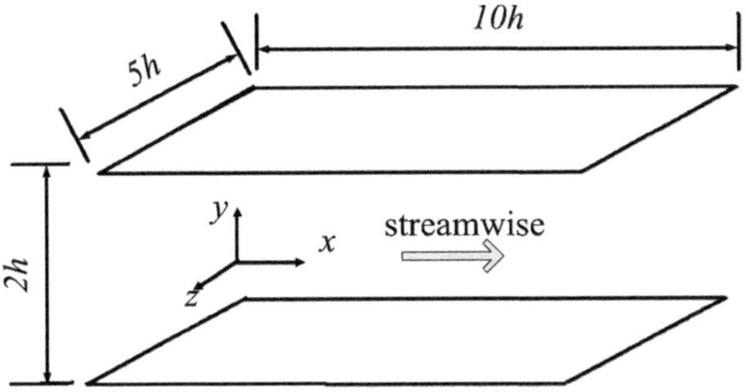

Fig. 1 Schematic of the simulated flow geometry, x, y and z represents the streamwise, wall-normal and spanwise directions, respectively, and u, v, w are the corresponding velocity components. In the streamwise, and spanwise directions, periodic conditions are imposed

The description of the characteristics of viscoelastic fluid was expressed by the well-known Giesekus model [20]. Besides, an additional consinusoidal force $\mathbf{F} = [F_m \sin(2\pi y/h),\ 0,\ 0]$ with F_m of the amplitude of the additional force was exerted to generate a strong rectilinear shear inside the channel and sustain the turbulent motion from the viscous and elastic dissipation. In the simulation, the dimensionless governing equations are expressed as follows:

$$\frac{\partial u_i}{\partial x_i} = 0 \tag{1}$$

$$\frac{\partial u_i}{\partial t} + u_j \frac{\partial u_i}{\partial x_j} = -\frac{\partial p}{\partial x_i} + \frac{1}{\mathrm{Re}} \frac{\partial^2 u_i}{\partial x_j \partial x_j} + \frac{\beta}{\mathrm{Re}Wi} \frac{\partial (C_{ij} - \delta_{ij})}{\partial x_j} + f_i^* \tag{2}$$

$$\frac{\partial C_{ij}}{\partial t} + u_k \frac{\partial C_{ij}}{\partial x_k} = C_{ik} \frac{\partial u_j}{\partial x_k} + C_{kj} \frac{\partial u_i}{\partial x_k} - \frac{(\delta_{ik} + \alpha (C_{ik} - \delta_{ik}))(C_{kj} - \delta_{kj})}{Wi} \tag{3}$$

Where, the subscript 'i' ($i, j = 1,2,3$ respectively, corresponding to x, y, z direction) satisfies the summation convention. The half height of the channel, h and the imaginary laminar velocity amplitude under the given additional force f_i, u_l, h/u_l and ρu_l^2 are chosen as scales for length, velocity, time, and stress, respectively, with $u_l = F_m h^2/(4\pi 2\eta_s)$, and η_s of the solvent contribution. β shows the solute (η_P) contribution to the apparent viscosity and is set to 0.5; $\mathbf{C}$ is the dimensionless conformation tensor showing the deformation of the embodied microstructures and normalized by η_P; $\mathbf{I}$ is the unit tensor; α is the mobility factor and set to be 0.001; $\mathrm{Re} = 2u_l h/(\eta_P + \eta_s)$, and $\mathrm{Wi} = 2\lambda u_l/h$, respectively, with λ of the polymers relaxation time; f^* is the dimensionless exerted force, $f_x^* = 2\pi\beta \sin(2\pi y)/\mathrm{Re}_l$. Besides, the global instantaneous kinetic energy and turbulent kinetic energy are defined as $E_k = \iiint_V (u^2 + v^2 + w^2)/2dV/V$ and $E_{kf} = \iiint_V \left(\overline{u'^2 + v'^2 + w'^2}\right)/2dV/V$.

In the numerical simulation, proper initial conditions should be imposed for velocity and conformation fields to induce the elastic turbulent state. In the present study, we used the numerical results obtained from the fully-developed elastic turbulent state of 3D viscoelastic Kolmogorov flow as the initial fields. All the numerical cases were conducted based on $64 \times 64 \times 64$ grids which were fine enough for a very low Re flow. Second-order central difference scheme and the MINMOD scheme were used for the spatial discretization of the convective term in momentum equations and Giesekus constitutive equations, respectively. For the time derivative term, Adams–Bashforth scheme was applied.

3 Results and Analysis

In the simulations the flow has been reached to the elastic turbulent state in the straight channel by identifying its energy spectrum with a decaying exponent of -3.8, and the result was consistent with the references [10, 15]. Firstly temporal evolutions of the global instantaneous and turbulent kinetic energies were shown in Fig. 2. It is found that the flow achieves a statistically fully-developed state with the relative turbulent intensity increasing with Wi to 5.2% at Wi $=$ 30. Moreover, the amplitude of flow oscillations becomes large and the intermittency appears with larger-time-scale flow structures generated as shown in the evolution of the E_{kf}.

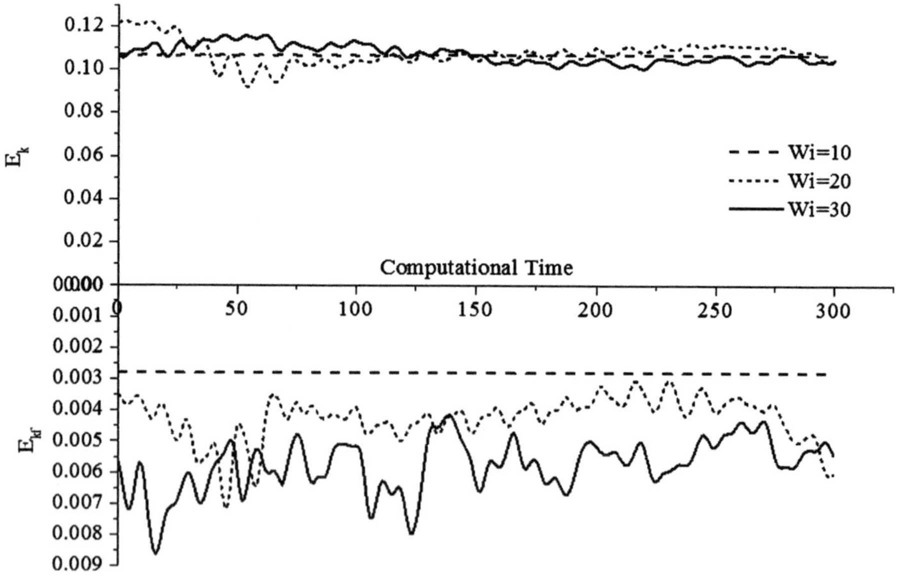

Fig. 2 Temporal evolution of the global kinetic energy E_k, and turbulent kinetic energy E_{kf} at different Wi

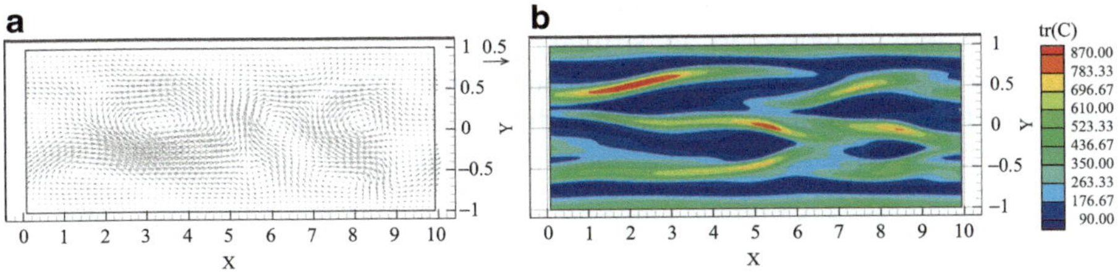

Fig. 3 The distribution of (**a**) the fluctuating velocity vector based on streamwise velocity *u'*, wall-normal velocity *v*, in which "0.5" with the arrow means the reference magnitude of the velocity vector and (**b**) polymers' elongation tr(**C**) in *x*-*y* plane at Wi = 30

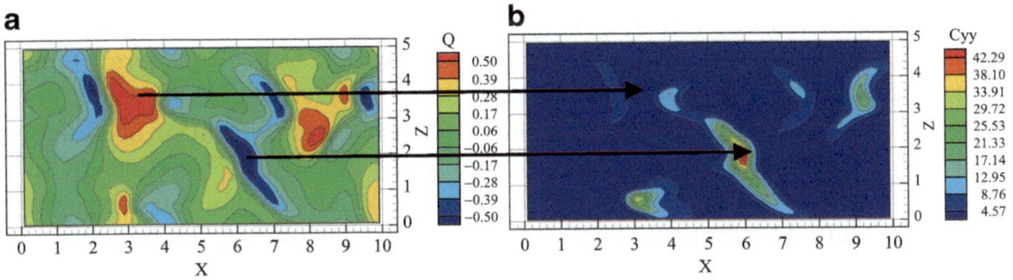

Fig. 4 The distribution of the vortex structures Q (**a**) and the polymers' elongation in the wall-normal direction Cyy (**b**) in x-z plane at Wi = 30. By comparing the two figures, the location of Q is closely related with the distribution of polymer's stretch in the wall-normal direction

This is attributed to the slow elastic-dissipation when the relaxation time of polymers increases.

Next, the snapshots of the flow patterns based on the streamwise fluctuating velocity ($u' = u - \langle u \rangle$) and the wall-normal one (v), and the polymers elongation ($\mathrm{tr(C)} = C_{xx} + C_{yy} + C_{zz}$) in the *x*-*y* plane are shown in Fig. 3, respectively, to get an intuitive interpretation of the flow under the elastic turbulent state. The main flow of velocity component *u* shows the periodicity in the normal-wise direction due to the additional cosine force. When the flow is in the elastic turbulent state, the velocity fluctuations *u'* and the wall-normal component *v* show some localized quadrupolar flow patterns with positive and negative values compared to the mean velocity. The localized vortex structures whose scale is of the order of the channel width are formed. In addition, the polymers' elongation shows the disordered filament structures which are embodied in the strong shear regions. Finally, the interactions between the flow structures and the polymer behaviors are briefly discussed. Figure 4 shows the relation between the vortex structures strength Q and the polymers' elongation in the wall-normal direction C_{yy}. Here, Q is the second invariant of the velocity-gradient tensor S ($S = (\nabla u + \nabla u^T)/2$), and indicates the strength of vortex tubes, $Q = W : W - S : S$ with $W = (\nabla u - \nabla u^T)/2$. From the figure, it is found that in the elastic turbulent vortex-tubes ($Q > 0$) are distorted in the spanwise direction, and closely accompanied by the deformed flow patterns ($Q < 0$), where the polymers are strongly stretched as shown in Fig. 4b. In other words, the vortex tubes are mainly formed inside the regions where the polymers are strongly stretched in the wall-normal direction: the polymers' stretching in this direction leads to the generation of the rotating motion.

4 Conclusions

In this paper, DNSs were carried out to study the characteristics of the velocity fields and the polymers' elongation in the elastic turbulent flow in the parallel plate channel: this means without a local curvature but with the wall effects. The following conclusions were drawn: (a) the relative turbulent intensity increases with increase of Wi, and the turbulent motion becomes stronger and more intermittent due to the slow elastic-dissipation under the larger Wi case; (b) the wall-normal velocity was induced due to the polymers' feedback in the form of quadrupolar patterns, and the scale of flow structures are of the order of the geometrical size; (c) the polymers' elongation shows the filament structures, surrounding which the vortex tube-like structures (regions dominated by rotating motion) tend to be formed.

Acknowledgment The authors were grateful for the support of the Ministry of Education, Culture, Sports, Science and Technology of Japan via "Energy Science in the Age of Global Warming" of Global Center of Excellence (G-COE) program (J-051).

References

1. Bird RB, Curtiss CF, Armstrong RC, Hassager O (1987) Dynamics of polymers liquids. Wiley, New York
2. Toms BA (1949) Some observation on the flow of linear polymer solutions through straight tubes at large Reynolds number. In: Proceedings of 1st international congress on Rheology, North Holland 2:135–141
3. Groisman A, Steinberg V (2001) Efficient mixing at low Reynolds numbers using polymer additives. Nature 410:905–908
4. Burghelea T, Segre E, Steinberg V (2004) Mixing by polymers: experimental test of decay regime of mixing. Phys Rev Lett 92:164501
5. Burghelea T, Segre E, Bar-Josephi I, Groisman A, Steinberg V (2004) Chaotic flow and efficient mixing in microchannel with polymer solutions. Phys Rev E 69:066305
6. Li FC, Kinoshita H, Li XB, Oishi M, Fujii T, Oshima M (2010) Creation of very-low-Reynolds-number chaotic fluid motion in microchannel using visocelastic surfactant solution. Exp Therm Fluid Sci 34:20–27
7. Groisman A, Steignburg V (2004) Elastic turbulence in curvilinear flows of polymer solutions. New J Phys 6:29
8. Groisman A, Steignburg V (2001) Stretching of polymers in a random three-dimensional flow. Phys Rev Lett 86:934–93

9. Burghelea T, Segre E, Steinberg V (2006) Role of elastic stress in statistical and scaling properties of elastic turbulence. Phys Rev Lett 96:214502
10. Burghelea T, Segre E, Steinberg V (2007) Elastic turbulence in von Karman swirling flow between two disks. Phys Fluids 19:053104
11. Thomas DG, Sureshkumar R, Khomami B (2006) Pattern formation in Taylor-Couette flow of dilute polymer solutions: dynamical simulations and mechanism. Phys Rev Lett 97:054501
12. Berti S, Bistagnino A, Boffetta G, Celani A, Musacchio S (2008) Two-dimensional elastic turbulence. Phys Rev E 77:055306
13. Berti S, Boffetta G (2010) Elastic waves and transition to elastic turbulence in a two-dimensional viscoelastic Kolmogorov flow. Phys Rev E 82:036314
14. Thomases B, Shelley MJ (2009) Transition to mixing and oscillations in a Stokesian visco-elastic flow. Phys Rev Lett 103:094501
15. Zhang HN, Li FC, Cao Y, Yang JC, Li XB, Cai WH (2011) The vortex structures of elastic turbulence in 3D Kolmogorov flow with polymer additives. In: Proceeding of the sixth international conference on fluid mechanics, Paper No. 153, Guangzhou, China, 30 June–3 July 2011
16. Li FC, Zhang HN, Cao Y, Kinoshita H, Oshima M, Kunugi T (2012) Purely elastic instability and mixing enhancement in a 3D curvilinear channel flow. Chin Phys Lett 29:094704
17. Wei JJ, Kawaguchi Y, Yu B, Feng ZP (2006) Rheological characteristics and turbulent friction drag and heat transfer reductions of a very dilute cationic surfactant solution. ASME J Heat Transfer 128:977–983
18. Larson RG (1992) Review: instabilities in viscoelastic flows. Rheologica Acta 31:213–263
19. Shaqfeh ESG (1996) Purely elastic instabilities in viscometric flows. Annu Rev Fluid Mech 28:129–185
20. Groisman A, Steignberg V (2000) Elastic turbulence in a polymer solution flow. Nature 405:53–55

Visualization Study on Bubbly Turbulent Flow in a Small Square Duct

HaoMin Sun, Tomoaki Kunugi, and DaZhuan Wu

Abstract Bubbly turbulent flow in a small square duct was visualized by a high-speed camera in various flow conditions. In order to find the appropriate method for generating bubbles with uniform size, two different bubble generators of the sintered metal with and without a plastic cone on its top were tested. The result showed that the bubble size using the sintered metal with the cone was more uniform compared to that without the cone. The bubble behavior depending on the bubble size was discussed for the bubble generation method with the plastic cone. It was found that the large bubble deformation in the vicinity to the wall in the small square duct was observed as same as that in a large duct.

Keywords Bubble generation method • Bubbly turbulent flow • Visualization

1 Introduction

In some energy utilities such as a light water nuclear reactor and a boiler, gas–liquid two-phase turbulent flows play a significant role from the viewpoint of both its safety and energy efficiency. Therefore, many studies regarding two-phase turbulent flows have been conducted so far in the circular pipe, such as Serizawa et al. [1, 2] and so on. In addition, in order to investigate the effect of a second kind of secondary flow, which was the well-known typical flow for the single phase turbulent flows in non-circular duct and was studied by Hoagland [3], on the bubbles, the authors have carried out the measurement of the local void fraction in the large square duct with 136 mm in the

H. Sun • T. Kunugi (✉)
Department of Nuclear Engineering, Kyoto University, Yoshida honmachi, Sakyoku, Kyoto 606-8501, Japan
e-mail: kunugi@nucleng.kyoto-u.ac.jp

D. Wu
Department of Chemical and Biological Engineering, Zhejiang University, Hangzhou, China

T. Yao (ed.), *Zero-Carbon Energy Kyoto 2012*, Green Energy and Technology,
DOI 10.1007/978-4-431-54264-3_31,

inner side length for various bubbly flow conditions [4]. Although it was found that the local void fraction showed the peak near the corner of the square duct, the effect of inlet flow condition was not discussed deeply. Since the inlet flow condition such as bubble size distribution strongly affects the void fraction distribution [5], bubbles with more uniform size in the specified flow condition are required in order to eliminate the effect of inlet flow condition on the void fraction distribution and reduce the experimental parameters to the superficial gas and liquid velocities only.

According to the importance of the bubble size uniformity mentioned above, the purpose of this study is to investigate the method for generating bubbles with more uniform size in the specified flow condition. The bubble size in the specified flow condition was discussed via the visualization experiments for bubbly turbulent flow in a small square duct. In addition, the bubble behavior in this small square duct was discussed by comparing with that in the large square duct reported by the authors [6].

2 Experimental Apparatus

A schematic diagram of the experimental apparatus is shown in Fig. 1. The small square acrylic duct with 24 mm in the inner duct side length was used as the test section. Tap water was stored in the storage tank and introduced to the test section by a centrifugal pump (MD-40R-N, Iwaki Corp., Japan). The water flow rate was measured by a turbine flow meter (LD20-RPATAAA, Horiba Corp., Japan) and controlled by a valve. Then the flow was divided into two lines, and connected to the lower water plenum of the duct. On the other hand, air was controlled by a mass flow meter (FLOWLINE SEF-5, Horiba Corp., Japan) and introduced through a stainless steel pipe into the sintered metal. Bubbles were generated by the mixing of water and air at the surface of the cylindrical sintered metal (SMC Corp., Japan) with the diameter and height of 8 and 13 mm respectively, as shown in the left hand side of Fig. 2. Vertical upward bubbly flow was formed in the test section at room temperature and atmospheric pressure. Sintered metals with two different pore sizes based on the nominal filtration density of 2 and 120 μm were used as the bubble generator. Different pore sizes didn't affect bubble behavior apparently.

Bubble behavior was visualized by a high-speed video camera (Fastcam-512PCI, Photron Corp., Japan) with the frame rate faster than 1,000 fps in both the sintered metal region and duct region which is 300 mm downstream from the sintered metal. Several flow conditions: the combination of the superficial liquid velocity $J_l = 0.5, 0.75, 1.0, 1.25$ m/s and the superficial gas velocity $J_g = 0.029$, 0.058, 0.087 m/s, were tested. All flow conditions were turbulent flow region because the minimum Re of the liquid phase based on the duct side length was more than 10,000.

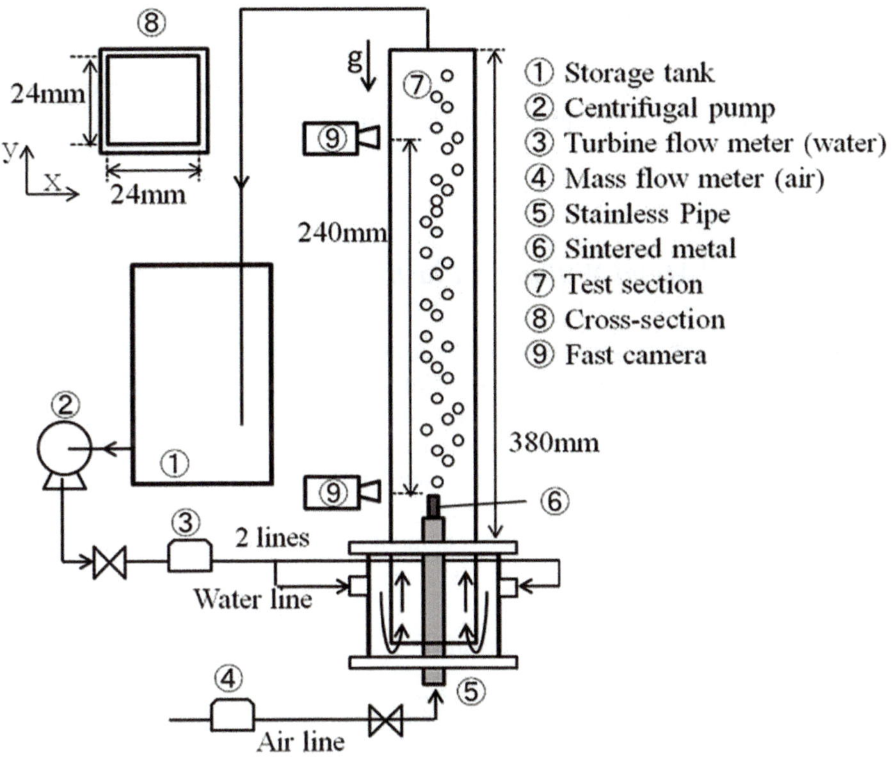

Fig. 1 Schematic diagram of experimental apparatus

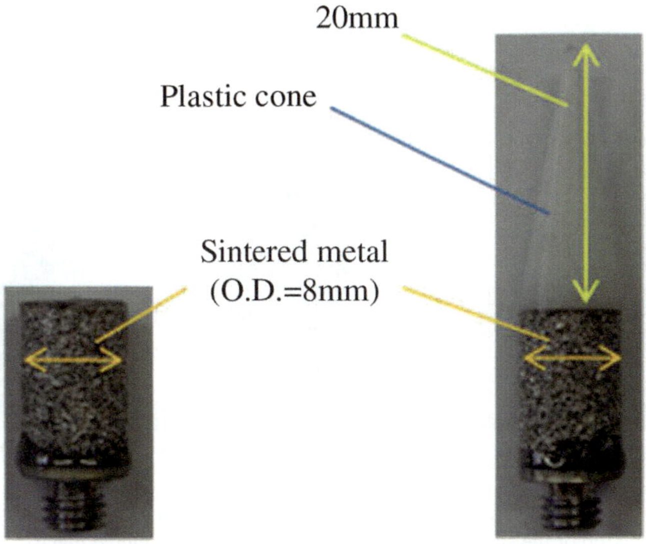

Fig. 2 Different types of bubble generators

3 Bubble Generation Method

In order to generate bubbles with uniform size, bubble behavior generated by using two different bubble generators: the sintered metal with and without the plastic cone attached on its top were visualized. The configurations are shown in Fig. 2.

3.1 Sintered Metal Without a Plastic Cone

First of all, the bubbles were generated by using the sintered metal without a plastic cone. Figure 3 shows the bubble behavior generated from the sintered metal region in various flow conditions of $J_g = 0.029$ m/s and $J_l = 0.5, 0.75, 1.0$ m/s respectively. Bubbles were observed to be dragged and accumulated on the top of the sintered metal. In addition, a bubble column was formed on the top of the sintered metal due to the bubble coalescence and enlarged its height with increase of the superficial liquid velocity. The surface shape of the bubble column waved with time especially in its top

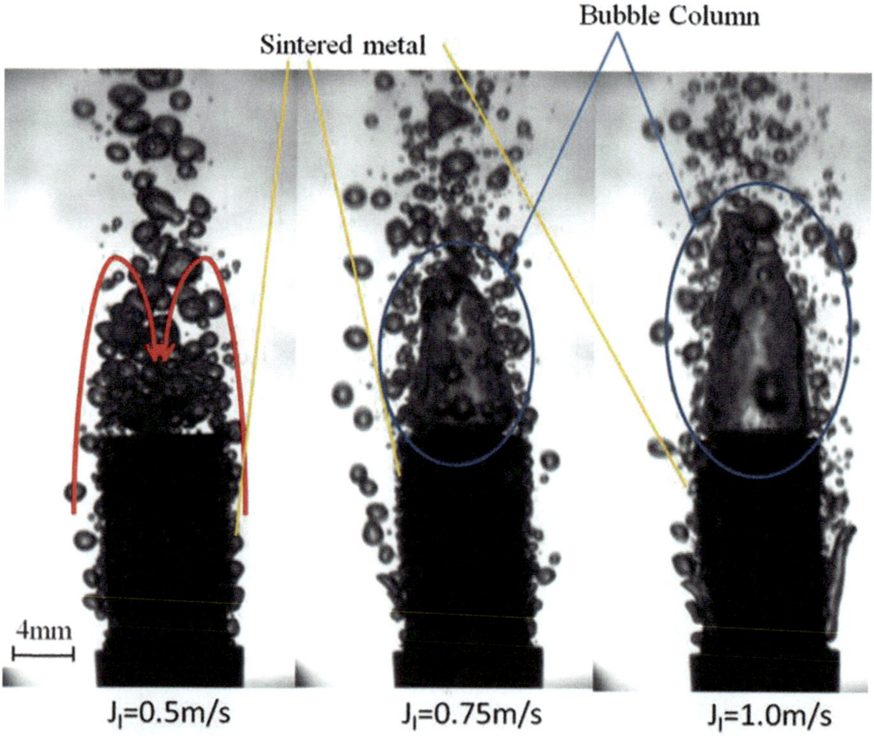

Fig. 3 Bubbles generated by sintered metal without a plastic cone, and a bubble column was appeared on the top of the sintered metal ($J_g = 0.029$ m/s, 2 μm sintered metal)

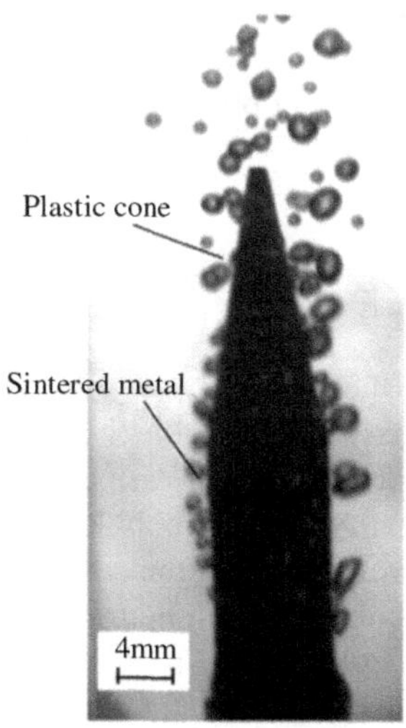

Fig. 4 Bubble behavior surrounding the sintered metal with the plastic cone

part. When the bubble column became long, the relatively large bubble was observed to be pinched off from the top of the bubble column. The size of such bubble was larger compared to the bubble generated at the surface of the sintered metal.

It is well known that when a single phase turbulent flow passes a circular cylinder, the flow separation occurs and a recirculating wake appears in the downstream behind the cylinder. In this experiment, the turbulent flow occurs near the sintered metal because the Re_d ($=J_l \times D/\nu$) based on the sintered metal diameter was more than 4,000. Here, D (8 mm) and ν are the outer diameter of the sintered metal and the kinematic viscosity of liquid phase. Therefore, it was considered that the bubbles interacted with the recirculating wake on the top of the sintered metal and were accumulated in the wake region. Since the length in the stream direction of the recirculating wake in single-phase turbulent flow is increased with increase of the Re_d, the height of a bubble column was enlarged with increase of the superficial liquid velocity.

3.2 Sintered Metal with a Plastic Cone

Since a large bubble pinched off from the bubble column was observed which causes non-uniform bubble size in the duct region, a plastic cone was attached on the top of the sintered metal in order to prevent a bubble column forming on the top of the sintered metal. Bubbles generated by the sintered metal with a plastic cone were also visualized in both the sintered metal region and the duct region. Figure 4 shows the bubbles in the sintered metal region under the flow condition of $J_l = 1.0$ m/s and $J_g = 0.029$ m/s. It was found that the bubble column was not formed due to the existence of the plastic cone. Since there was no bubble column on the top of the sintered metal, the relatively large bubbles pinched off from the bubble column were not observed in the duct region.

4 Visualization of Bubble Behavior

Bubble behavior generated by the sintered metal with the plastic cone was visualized. Table 1 shows the bubbles in both the sintered metal region and the duct region in various flow conditions.

As for the bubble size, it was found that the increase of the superficial liquid velocity with the same superficial gas velocity resulted in decreasing the average bubble size in both regions. In addition, the bubble size was not changed so much between two regions. It was suggested that the bubble size in the duct region depended on the flow condition in the sintered region rather than the bubble coalescence during their rising.

As for concerning the shape of bubbles, it was observed that the bubbles in vicinity of the wall tend to have a sharp angle toward the wall especially for the small superficial liquid velocity condition shown in Table 1. It was also reported in the large square duct by the authors [6]. The interaction between the bubble and the vortexes generated in the liquid phase near the wall could be very important to explain the large bubble deformation because of the vortexes exist in this turbulent flow condition (minimum Re of the liquid phase based on the duct inner side length was more than 10,000).

Table 1 Bubble behaviors in different flow condition

Flow Condition	J_l=0.5m/s, J_g=0.087m/s		J_l=1.0m/s, J_g=0.087m/s		J_l=1.25m/s, J_g=0.087m/s	
Visualization Region	Sintered Metal	Duct	Sintered Metal	Duct	Sintered Metal	Duct

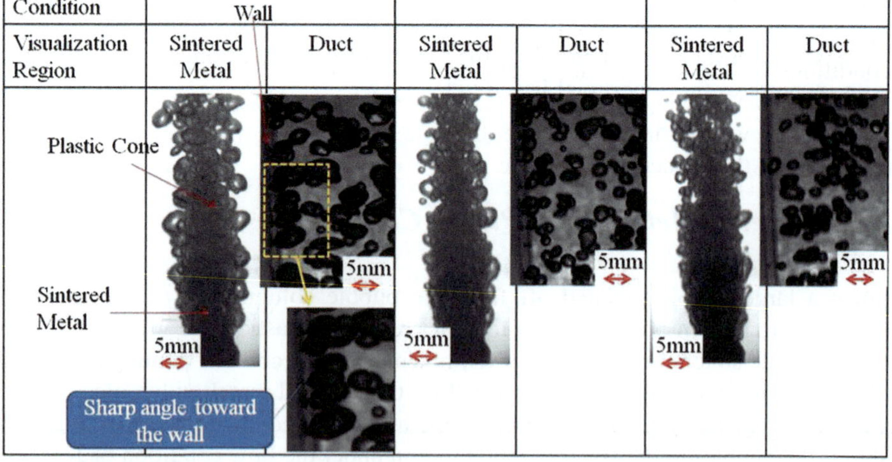

5 Concluding Remarks

1. The plastic cone attached on the sintered metal was enhanced the uniformity of the bubble size by preventing the bubble column formation on the top of the sintered metal.
2. The large bubble deformation was found in the small square duct as same as that in the large square duct.

Acknowledgement This work was supported by the Aid for the Japan Society for the Promotion of Science (JSPS) research fellows. In addition, the authors were grateful for "Energy Science in the Age of Global Warming" of Global Center of Excellence (G-COE) program (J-051) of the Ministry of Education, Culture, Sports, Science and Technology of Japan.

References

1. Serizawa A, Kataoka I, Michiyoshi I (1975) Turbulence structure of air-water bubbly flow – 2. Local properties. Int J Multiphase Flow 2:235–246
2. Serizawa A, Kataoka I (1990) Turbulence suppression in bubbly two-phase flow. Nucl Eng Des 122:1–16
3. Hoagland LL (1960) Fully developed turbulent flow in straight rectangular ducts – secondary flow, its cause and effect on the primary flow. Ph.D. thesis, Department of Mechanical Engineering, Massachusetts Institute of Technology
4. Sun HM, Kunugi T (2012) Gas–liquid two-phase turbulent flow in square duct, Zero-carbon energy Kyoto 2011. Green Energy Technol III (iii):279–285
5. Liu TJ, Bankoff SG (1993) Structure of air-water bubbly flow in a vertical pipe-2. Void fraction, bubble velocity and bubble size distribution. Int J Multiphase Flow 36(4):1061–1072
6. Sun HM, Kunugi T, Wu DZ, Zhang HN, Nakamura H, Shen XZ (2012) Gas–liquid bubbly turbulent upward flow in square duct. In: The 20th international conference on nuclear engineering, Anaheim, USA, 30 July–3 August 2012. ICONE20POWER2012-54918, proceeding, accepted

A Feasibility Study of a Medical Application of Cylindrical Discharge Type Fusion Neutron Beam Source

Yasunori Nakai, Kazuyuki Noborio, Ryuta Kasada, Yasushi Yamamoto, and Satoshi Konishi

Abstract The present study aims to investigate the feasibility of a cylindrical type electric discharge nuclear fusion neutron beam for Boron Neutron Capture Therapy (BNCT). The MCNP5 code estimates the neutron transportation in a human brain phantom which consists of the skull (Ca), the brain (water), and the tumor affected part (^{10}B). Monochromatic neutron sources (10^6 neutrons per second) with incidence of arbitrary energy spectra (14 MeV, 2.45 MeV, 100 keV, 10 keV, 1 keV, 100 eV, 10 eV, 1 eV, 0.1 eV, 0.01 eV, monochromatic each) are examined. The result indicates that to the 8 cm-deep affected part, epithermal neutron with the energy near 10 keV is suggested to be effective to penetrate and maximum dose by the capture reaction by boron in medical treatment.

Keywords Beam • BNCT • MCNP • Medical application • Nuclear fusion neutron source

1 Introduction

Boron Neutron Capture Therapy (BNCT) [1] is a radiation therapy for cancer using a neutron beam and a ^{10}B-loaded drug to kill tumor cells selectively. Figure 1 shows simplified schematic of BNCT.

Y. Nakai (✉)
Department of Conversion Energy Science, Graduate School of Energy Science, Kyoto University, Gokasho, Uji 611-0011, Japan
e-mail: yasnakai@iae.kyoto-u.ac.jp

K. Noborio • R. Kasada • S. Konishi
Institute of Advanced Energy, Kyoto University, Gokasho, Uji 611-0011, Japan

Y. Yamamoto
Faculty of System Engineering, Kansai University, Yamate-cho 3-3-35, Suita 564-8680, Japan

T. Yao (ed.), *Zero-Carbon Energy Kyoto 2012*, Green Energy and Technology,
DOI 10.1007/978-4-431-54264-3_32, © Springer Japan 2013

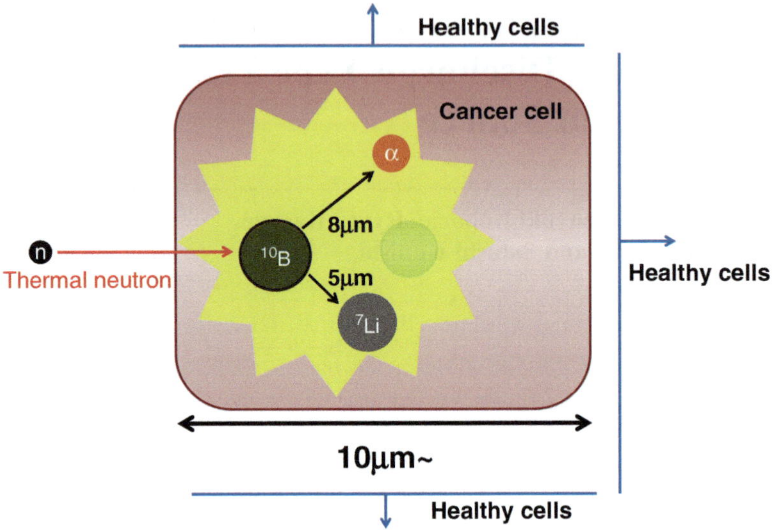

Fig. 1 Simplified schematic of BNCT

The principle of BNCT is as follows: (I) ^{10}B atoms are sent into a cancer cell target selectively by the difference with healthy cells in accumulation nature [2] or Drag Delivery System [3]. (II) Thermal neutrons are irradiated to induce ^{10}B (neutron, alpha, gamma)^{7}Li reaction. (III) Alpha particles (about 8 μm of range), ^{7}Li particles (about 5 μm of range), and the gamma ray of 478 keV are generated at this reaction. (IV) Since the diameter of a cell is generally 10–15 μm, the reaction ends within a cell. (V) As a result, a cancer cell becomes extinct selectively.

While fission reactors have been used to generate neutron beam for the BNCT [4], availability of neutron beam source is one of the important issues to extend the opportunity of BNCT. A cylindrical discharge tube type fusion neutron beam source is attractive for the BNCT because of its advantages: the size of devices is enough small to be installed into hospitals; suitable set-up of reflector and moderator (Fig. 2) can generate neutrons having selective energy spectra [5]. This device makes fusion reaction on the surface of cylindrical electrode (Fig. 3) with titanium vapor coating, by the glow discharge of hydrogen isotopes (deuterium or tritium) [6, 7]. And it generates a monochromatic fast neutron by the fusion reaction by the collision of the ion accelerated by high-voltage electric field. The neutron beam slowed down arbitrarily can be generated by covering the circumference by the reflector of a cylinder type which released the end. Therefore, this device makes it possible to supply neutron which has arbitrary energy spectra.

There have no precedent to have examined the possibility aiming at hospital establishment-type medical treatment for BNCT with the cylindrical discharge type fusion device itself in the world. Furthermore, until now the patient was decided according to the neutron spectrum which fission reactors can take out. However, the

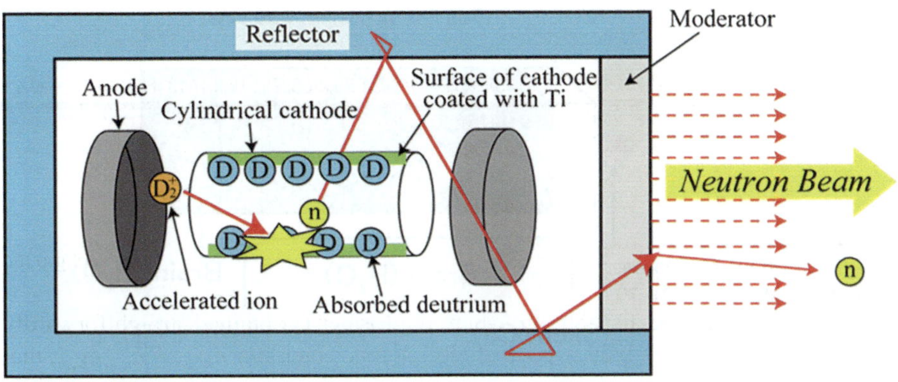

Fig. 2 Schematic representation of cylindrical discharge type fusion neutron beam source

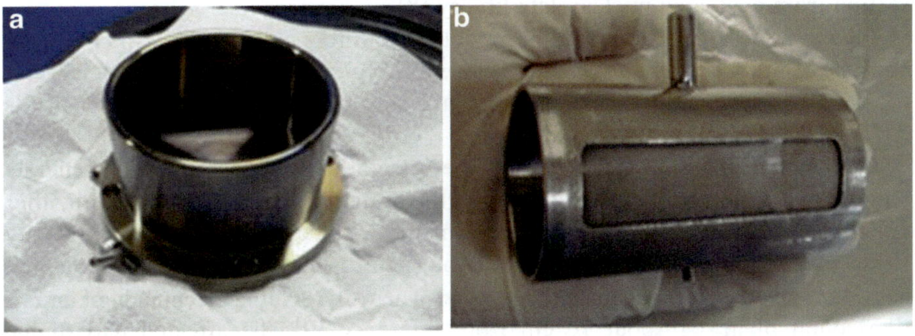

Fig. 3 Pictures of electrodes: (**a**) anode, (**b**) cathode

concept which aims at adaptation in broader case by choosing neutron spectrum arbitrarily according to a patient does not look at an example.

In this study, we have calculated the energy spectra in a simplified head phantom irradiated with monochromatic neutron beam by use of neutron transportation analysis in order to check the feasibility of BNCT using the cylindrical discharge tube type fusion neutron beam source.

Although it cannot say that medical-engineering collaboration is still enough even now, this research could be contributed to development of the collaboration.

2 Method

Monte Carlo Neutron Particle calculation code version5 (MCNP5) was used for neutron transport analysis. The MCNP5 is one of Monte Carlo method calculation codes, and it calculates the action of a neutron flux by random numbers in

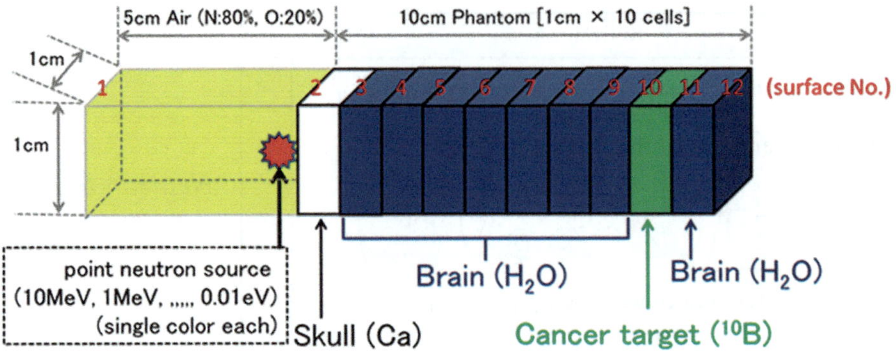

Fig. 4 MCNP5 phantom calculation model

accordance with Boltzmann neutron transport equation and nuclear reaction cross-section from evaluated nuclear data library [8].

Figure 4 shows a head phantom model used in this calculation. This phantom consists of the skull which is Ca (density is 1.55 [g/cm^3]), the brain which is water (density is 1.0 [g/cm^3]), and the tumor affected part containing ^{10}B (density is 2.07 [g/cm^3]) where it fixes to a depth of 8 cm.

On the characteristic, the common experimental fission reactors which generate thermal neutron have been treated only to the brain tumor of the 3 cm- depth, after opening the skull. 5 cm was a limit from the surface also in Kyoto University Reactor (KUR) which generates epithermal neutron improved so that it was not necessary to open a skull in recent years. Then, the average diameter of brains (horizontal plane) is about 16 cm from reference [9], the affected part of the phantom was assumed in the 8 cm-depth which is the center of the brain in order to challenge the limit of medical treatment and the aim of the tumor which occurs in the central part of brain such as pineal tumor, choroid plexus tumor, etc.

The cells are divided for every cm. The attenuation, thermalization and absorption in the phantom were evaluated.

A point neutron source at 10^6 neutrons per second is set 0.05 cm near from surface number 2 and supposed with incidence of arbitrary monochromatic energies: 14 MeV, 2.45 MeV, 100 keV, 10 keV, 1 keV, 100 eV, 10 eV, 1 eV, 0.1 eV, 0.01 eV. An exposure field is made into 1 cm around. This system is enough to evaluate whether neutrons reach to affected part as 1 cm cube being in the back of collimator with 1 cm around as assumption.

3 Results and Discussions

Figure 5 shows neutron attenuation at the time of each cells passage calculated by MCNP5. It is found that neutrons of 100 keV and more pass the affected part at a depth of 8 cm containing ^{10}B and neutrons of less than 1 keV do not reach to the

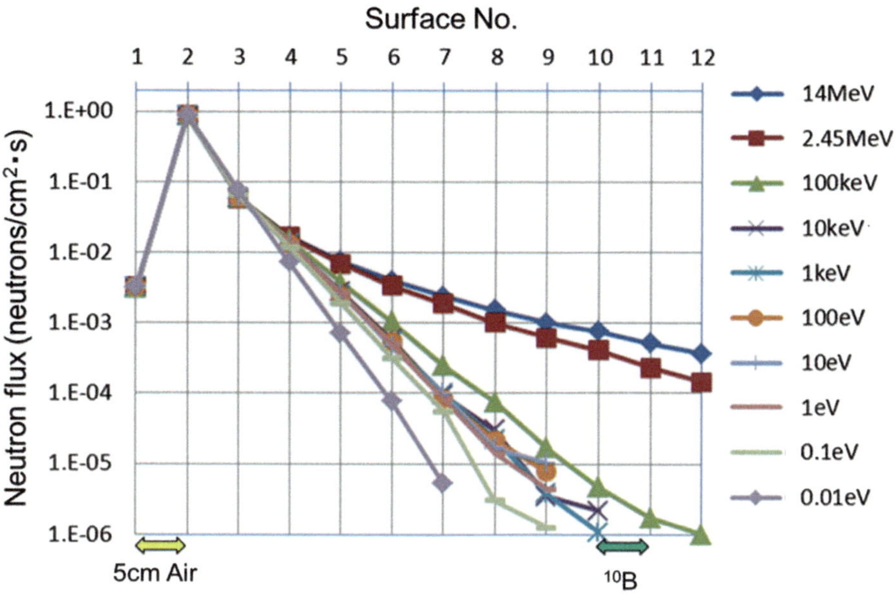

Fig. 5 Neutron attenuation at the time of each cell passage

affected part. Since the program returned the neutron flux value of zero, it was regarded that it disappeared.

About attenuation, epithermal and thermal neutrons decrease at a about same rate to the last, but inclination of fast neutrons decrease become loose suddenly when they pass surface 4. About thermalization on surface 10, as opposed to the whole flux of 7.2×10^{-4}[neutrons/cm^{2}s], almost all (6.7×10^{-4}[neutrons/cm^{2}s] as 94%) 14 MeV neutrons do not slow down, while 10 keV neutrons slow down (as whole flux of 1.6×10^{-6}[neutrons/cm^{2}s] versus energy range abundance ratio is 10 keV ~ 1 keV : 3.1×10^{-7}[neutrons/cm^{2}s] as 19%, 1 keV ~ 100 eV : 5.3×10^{-7}[neutrons/cm^{2}s] as 34%, 100 eV ~ 10 eV : 2.6×10^{-7}[neutrons/cm^{2}s] as 17%, 10 eV ~ 1 eV: 4.8×10^{-4}[neutrons/cm^{2}s] as 30%, respectively). Figure 6 shows cross section of ^{10}B from Japanese Evaluated Nuclear Data Library (JENDL) 4.0 [10]. According to Fig. 6, cross section of (n,α) reaction on 14 MeV is 48.95[mbarn], so (n,α) reaction probability is astronomically low. On 10 keV, if average value is calculated in consideration of the rate of an energy range abundance ratio and cross section linearity, it is 238[barn]. Therefore, the probability that the neutron which reached ^{10}B cell will cause a (n,α) reaction is 5-figure high compared with the time of 14 MeV. By the same token, if (n,α) reaction with neutron absorption probability is calculated from cross section, the appropriate neutron absorption happens for the energy at around 10 keV.

This is a spectrum range which cannot be selected out at fission reactors well but can be selected out only with this equipment very well. These results suggest that the examination of the medical treatment using this equipment is effective.

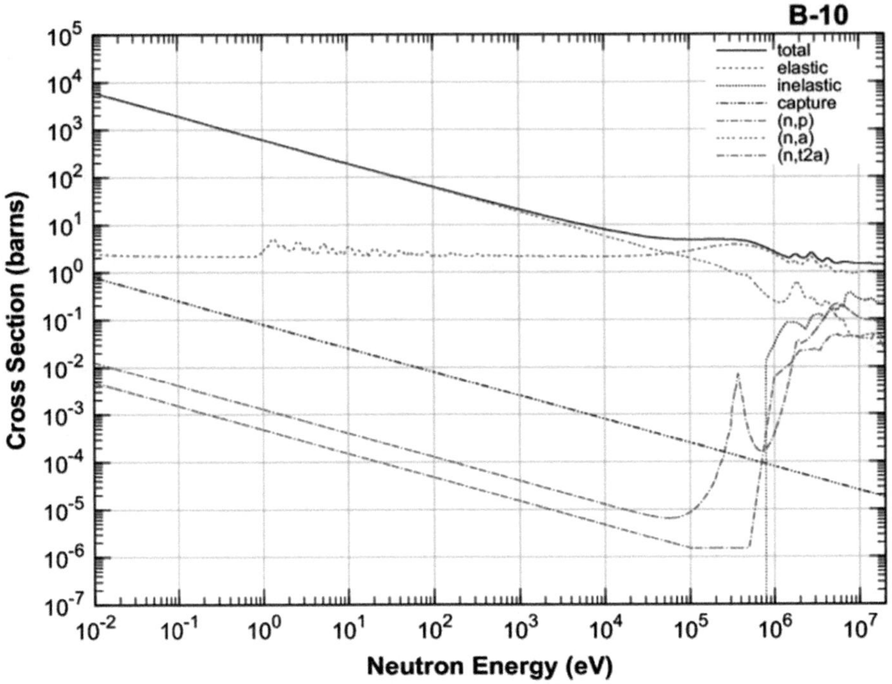

Fig. 6 Cross section of ^{10}B from JENDL 4.0

Further calculation on the detailed phantom model is necessary to investigate.

In the future, furthermore, we must make medical irradiation plans not to be focused on neutron source but patient's condition and quality of life. This research leads to realize order-made neutron irradiation medical treatment for patients.

4 Conclusion

The suitable spectrum for the BNCT medical treatment in the 8 cm-deep affected part is clarified. Epithermal neutron with the energy near 10 keV is suggested to be effective to penetrate and cause maximum dose by the capture reaction by boron in medical irradiation treatment. This is the spectrum range which cannot be taken out at fission reactors selectively well but can be taken out only with our cylindrical type electric discharge nuclear fusion neutron beam source with high selectivity. These results suggest that the examination of the medical treatment using this equipment is effective.

It is summarized that a feasibility of a medical application of cylindrical discharge type fusion neutron beam source is opened at future.

Acknowledgment This work was partly supported by an "Energy Science in the Age of Global Warming" of Global Center of Excellence (G-COE) program (J-051) of the Ministry of Education, Culture, Sports, Science and Technology of Japan.

References

1. Suzuki M et al (2007) Current and future of boron neutron capture therapy. Med Online 39 (12):1126–1135
2. Yoshino K (2011) Energy Rev 11:15–16
3. Nakamura H (2010) Minimally invasive cytoselective radiation therapy using boron neutron capture reaction. Yakugaku Zasshi 130(12):1687–1694
4. Nakagawa Y et al (2003) Clinical review of the Japanese experience with boron neutron capture therapy and a proposed strategy using epithermal neutron beams. J Neuro Oncol 62:87–99
5. Fatemeh SR et al (2012) Design of model for BSA to meet free beam parameters for BNCT based on multiplier system for D-T neutron source. Ann Nucl Energy 39:18–25
6. Kanagae T (2010) entou houdengata kakuyuugou chuuseishigen wo mochiita chuuseishi beam no neutronics kenkyuu (in Japanese) Master Thesis, Graduate School of Energy Science, Kyoto University
7. Maegawa T (2012) kakuyuugou chuuseishi beam-gen no neutronics to houdentokusei ni kansuru kenkyuu (in Japanese) Master Thesis, Graduate School of Energy Science, Kyoto University
8. Los Alamos National Laboratory X-5 Monte Carlo Team (2010) MCNP-A General Monte Carlo N-Particle Transport Code, Version 5
9. Shinohara H et al (1993) Attenuation correction for SPECT of regional brain perfusion using approximate thickness of skull-computer simulation study. Med Image Technol 11 (2):133–142
10. Shibata K et al (2011) JENDL-4.0: a new library for nuclear science and engineering. J Nucl Sci Technol 48(1):1–30

Author Index

T. Yao (ed.), *Zero-Carbon Energy Kyoto 2012*, Green Energy and Technology, DOI 10.1007/978-4-431-54264-3, © Springer Japan 2013

Keyword Index

T. Yao (ed.), *Zero-Carbon Energy Kyoto 2012*, Green Energy and Technology,
DOI 10.1007/978-4-431-54264-3, © Springer Japan 2013

GPSR Compliance

The European Union's (EU) General Product Safety Regulation (GPSR) is a set of rules that requires consumer products to be safe and our obligations to ensure this.

If you have any concerns about our products, you can contact us on ProductSafety@springernature.com

In case Publisher is established outside the EU, the EU authorized representative is:

Springer Nature Customer Service Center GmbH
Europaplatz 3
69115 Heidelberg, Germany

Batch number: 10755227

Printed by Printforce, the Netherlands